Quality of Life, Environmental Changes and Subjectivity

Sônia Regina da Cal Seixas
João Luiz de Moraes Hoefel

Quality of Life, Environmental Changes and Subjectivity

A Contribution to a New Line of Research on Climate Change

Sônia Regina da Cal Seixas
University of Campinas
Campinas, São Paulo, Brazil

João Luiz de Moraes Hoefel
UNIFAAT University Center
Atibaia, São Paulo, Brazil

ISBN 978-3-030-39221-5   ISBN 978-3-030-39222-2 (eBook)
https://doi.org/10.1007/978-3-030-39222-2

© The Editor(s) (if applicable) and The Author(s), under exclusive licence to Springer Nature Switzerland AG 2022
This work is subject to copyright. All rights are solely and exclusively licensed by the Publisher, whether the whole or part of the material is concerned, specifically the rights of translation, reprinting, reuse of illustrations, recitation, broadcasting, reproduction on microfilms or in any other physical way, and transmission or information storage and retrieval, electronic adaptation, computer software, or by similar or dissimilar methodology now known or hereafter developed.
The use of general descriptive names, registered names, trademarks, service marks, etc. in this publication does not imply, even in the absence of a specific statement, that such names are exempt from the relevant protective laws and regulations and therefore free for general use.
The publisher, the authors and the editors are safe to assume that the advice and information in this book are believed to be true and accurate at the date of publication. Neither the publisher nor the authors or the editors give a warranty, expressed or implied, with respect to the material contained herein or for any errors or omissions that may have been made. The publisher remains neutral with regard to jurisdictional claims in published maps and institutional affiliations.

Cover pattern © Melisa Hasan

This Palgrave Macmillan imprint is published by the registered company Springer Nature Switzerland AG.
The registered company address is: Gewerbestrasse 11, 6330 Cham, Switzerland

To my father José de Assis Seixas (in memorium) for his example.
To my sons Thiago, Samuel, Pedro, my daughters-in-law Maite and Juliana, and my son-in-law Daniel, for our family and for the meaning of life
To my grandchildren Miguel, Arthur, Catharina, Rafael and Cecilia, for the hope of the future
Sônia Regina
To my partner David S. Ferreira for his continuous support
João Luiz

Preface

This book has been organized into four chapters:

Chapter 1 is both an introduction and outline of what led us to write this text. We discuss what *Quality of Life (QoL)* means, and the implications that environmental change may potentially have on Quality of Life.

In Chap. 2 we present the concept of *subjectivity* and different approaches and contributions to the topics of *environmental change* and *climate change*. We also present the theoretical elements that have guided us and the role that these different fundamental disciplines contribute to the understanding of the concept of subjectivity. Psychoanalysis is one of the disciplines that play an important role in understanding subjectivity in relation to humanity's ability to face climate change and the enormous challenges arising from extreme events that are already occurring on a planetary level. One obvious example of an extreme planetary event is the Covid-19 pandemic. Covid-19 is not only a tragedy in itself, but also an example of what can happen on the planet if we fail to understand how to deal with climate and environmental changes.

In Chap. 3 we seek to present a critical and analytical, authorial approach to the challenges presented in the preceding chapter. These include the *Declaration of Human Rights (UN, 1948) and Agenda 2030 (UN 2015)*. In this chapter, we describe in detail the heart of this project. We elaborate on the contribution of psychoanalytical approaches for understanding and effectively contributing to the subjective actions of humanity. And, we bring awareness to the enormity of challenges such as climate change, extreme events, pandemics and the globalization of social environmental problems. These monumental problems can only be overcome, mitigated, or treated through a powerful process of adaptation that recognizes and encourages humanity to join collectively in creating a society with a *Common Future*. Only through *Global Agendas* will it be possible to overcome these challenges and the immense political and social obstacles that lie ahead and not fall into the barbarity that is the fruit of social inequality. This perspective emerges as a new theoretical and methodological consolidation that contributes to this reflection and to meeting the future challenges of the contemporary world.

In Chap. 4 – Conclusion, we bring up the most fundamental points with which we seek to analyse the fundamental problem presented in this book. We nurture the idea of subjectivity and its different approaches and contributions in the context of environmental and climate change, to help improve the quality of life, and consequently defend the dignity of life.

São Paulo, Brazil   Sônia Regina da Cal Seixas
Winter 2021   João Luiz de Moraes Hoefel

Acknowledgements

The authors are extremely grateful to the Center for Environmental Studies and Research (NEPAM) of Campinas University (UNICAMP) and the Centre for Sustainability and Cultural Studies (NESC), UNIFAAT – University Center.

To the Laboratory of Studies Environmental Changes, Quality of Life and Subjectivity – LEMAS/CNPq/NEPAM, for being an essential space for reflection and debate.

To our students from the doctoral programme Environment and Society (NEPAM/IFCH/UNICAMP) and post-graduate programme Energy Systems Planning (PSE/FEM/UNICAMP).

To the São Paulo State Research Support Foundation - FAPESP, for the financial support we have received in the last decade from especially the processes 2012/22578-1; 2013/17173-5; 2016/18585-3; 2019/08044-3, and to CNPq for the productivity scholarship to the first author.

We also thank our colleague Gabriela Farias Asmus, from the Federal University of ABC Paulista (UFABC), São Paulo, Brazil, for her critical reading of the first version.

To Debra Barbarick and Deborah Nickel from Sierra Water Protectors, Auburn, California, USA, for the translation support.

Contents

1 **Introduction** 1
 1.1 *Initial Approach: The Importance of This Subject and the Construction of New Paths* 2
 1.2 *Quality of Life and Environmental Changes* 4
  1.2.1 *Quality of Life Concept* 4
  1.2.2 *Quality of Life in a Climate Change Context* 6
 References 12

2 **The Importance of the Concept of Subjectivity: New Lines of Research** 15
 2.1 *Approaches* 16
 2.2 *Disciplinary Dimensions: Expanding Views and Concepts* 18
 2.3 *Climate Change and Subjectivity* 20
 2.4 *The Construction of Climate Change Narratives: 'Nightmare' or 'False Comfort'* 22
 2.5 *Climate Change and Psychoanalysis* 27
  2.5.1 *The Cultural Narrative About Climate Change and Subjectivity: Important Findings in Convergent Approaches* 30
 2.6 *Some Important Considerations About Objective Advances in the Relationship of Subjectivity, Psychoanalysis, and Climate Change. The Construction of Networks* 54
 References 63

xi

3 Building New Perspectives and Approaches to Our Common Future on Climate Change and Subjectivity: Agenda 2030 and Human Rights 69
   References 81

4 Conclusion 83
   Reference 88

Index 89

ABOUT THE AUTHORS

Sônia Regina da Cal Seixas has a PhD in social sciences from Campinas State University, UNICAMP, Brazil and Post-Doctoral in University of Reading/UK. Nowadays she is researcher at the Center for Environmental Studies and Research, NEPAM, State University of Campinas, UNICAMP. She is professor of doctoral programme Environment and Society (NEPAM-IFCH-UNICAMP) and of post-graduate programme Energy Systems Planning (FEM-UNICAM). She has a productivity research fellowship CNPq – Brazil's National Council for Scientific and Technological Development and Group Leader-Directory CNPq: Lab of Environmental changes studies, Quality of life and subjectivity. http://lattes.cnpq.br/4762940910820774

João Luiz de Moraes Hoefel has a PhD in social sciences at Campinas State University (IFCH/UNICAMP), São Paulo, Brazil, with a focus on environmental issues and developed post-doctoral research at the Center for Environmental Education and Public Policies at São Paulo University (ESALQ/USP). He teaches at UNIFAAT University Center, Atibaia, São Paulo, where he also coordinates the Center for Sustainability and Cultural Studies (NESC/CEPE/FAAT) developing research projects on natural resources conservation, environmental education and environmental planning. He also develops research activities at NEPAM/UNICAMP on global environmental change and quality of life as a collaborator researcher. http://lattes.cnpq.br/7635072427530391

CHAPTER 1

Introduction

Abstract This book is the fruit of two decades of intense reflection about sustainability, environmental change and subjectivity. This line of thought has brought about extensive research, which has resulted in the training and development of undergraduate and post-graduate students who are focused on these issues. This chapter seeks to highlight three important aspects: Initial Approach: The importance of this subject and the construction of new paths, Quality of Life and environmental changes, and Quality of Life Concept and Quality of life in a climate change context.

Keywords Sustainability · Environmental change · Subjectivity · Quality of life · Climate change

> What I learned over these decades is that we need an awakening, because if once upon a time it was, we indigenous peoples that were threatened with rupture or extinction of the meaning of our lives, today we are all faced with the reality that Earth can longer support our demands. (Krenak, 2020: 45[1])

---

[1] Krenak, A. (2020). *Ideias para adiar o fim do mundo*. São Paulo: Companhia das Letras.

© The Author(s), under exclusive license to Springer Nature Switzerland AG 2022
S. R. da Cal Seixas, J. L. de Moraes Hoefel, *Quality of Life, Environmental Changes and Subjectivity*,
https://doi.org/10.1007/978-3-030-39222-2_1

1.1 Initial Approach: The Importance of This Subject and the Construction of New Paths

This book is the fruit of two decades of intense reflection about sustainability, environmental change, and subjectivity. This line of thought has brought about extensive research, which has resulted in the training and development of undergraduate and post-graduate students who are focused on these issues. When Palgrave Macmillan invited us to publish this book, we were working on establishing a backdrop for research on the role of women in building sustainability and guaranteeing human rights.

For some very special reasons this invitation arrived at the right time. First of all, since 2018 Brazilian scientists, teachers, and professors have been living through the worst period in Brazilian history since the democratisation of the country, that followed the civil military dictatorship. Also, since 2019 there has been a systematic advance against the importance and legitimacy of the role played by science and scientists. There has also been a powerful attack on public policies regarding essential human rights previously achieved through tremendous efforts, by civil society.

In this context, our focus is centred on three supporting pillars:

1. To show the role of science in the context of social, cultural, and environmental development.
2. To recognize the anthropogenic (human caused) role in recent environmental change, with special emphasis on climate change.
3. To acknowledge the importance of an interdisciplinary approach to bring about positive and effective answers to the question of climate change, and the analysis of the impacts of human subjectivity. The psychoanalysis of human subjectivity offers a new approach. The inclusion of this perspective offers insight into the formation of support networks, the spread of science, and the forming of professional groups. Putting these conduits in place will create new and subjective ways of coping with climate change.

After its dire beginning in the middle of December 2019 in Wuhan, China, the World Health Organization declared (March 11, 2020) that Covid-19 was a pandemic. Producing this book in this climate compels us to pay attention to the importance of what happened in 2020. The pandemic proceeded to rapidly spread to the rest of the world, arriving in the Americas within just two months. The pandemic forced social isolation, physical distancing, the use of face masks, increased awareness of hygiene, and the closing of frontiers to potential travellers. The Covid-19 pandemic imposed a global effort on the world to find an adequate vaccine which was the only way to contain the pandemic as no other forms of early treatment or medicines existed to cure the disease. Through their efforts, scientists were able to achieve, in record time, a considerable number of vaccines and in the last quarter of 2020 we see many countries successfully initiating vaccination of their populations.[2]

Another important piece to consider, is the grave economic crisis that a majority of countries were already facing prior to the pandemic. The pandemic further aggravated the economic crisis, which amplified poverty and social inequality. Countries were already challenged by impoverished natural resources, deforestation, and loss of rights of numerous populations – especially the most vulnerable (First Nations people, traditional populations, quilombos, LGBTQIA+ people,[3] women, and the Black community, etc.).

This book is largely based on the ideas already mentioned, but we mustn't neglect to point out two fundamental things:

1. Our platform as engaged social scientists is to defend the dignity of life and to build a safe environment for everyone.
2. Our reflections are associated with, and have been guided by, two fundamental Agendas for society and its civilizational process: The Universal Declaration of Human Rights (UN, 1948) and Agenda 2030 (UN, 2015).

Without these two civilizing agreements it would be very difficult to face climate change, recognize the anthropogenic actions causing these events, and address the need to eradicate poverty and inequality in the world. Recognizing the importance of science and scientists is fundamental to achieving real goals that address these and other difficulties facing the

---

[2] In Brazil, the first two vaccines Coronavac-Butantan/SP and Astrazena/Oxford/Fiobruz/Br, slowly began to be given to priority groups, the elderly and the health professional population on the front lines of the fight against the pandemic in the middle of January and February 2021.

[3] LGBTQIA + refers to sexual and gender orientation, where each of the letters means lesbian, gay, bisexual, transgender, transsexual and transvestite, Queer, Intersexual, Asexual and the symbol +, are all the numerous other possibilities of sexual orientation and identity of gender.

modern world. Without this recognition we will not be able to build the agreements and solidarity needed by future generations.

In this context the role of *subjectivity* is fundamental. We recognize that each individual now faces, and will continue to face, the challenges we speak of here. Individuals will experience these challenges in unique ways; not only as a result of the objective conditions of their lives, but also in terms of their subjectivity and the framework we each have in place for facing challenging events that arise in our lives. From this perspective we see that the dynamic of subjectivity is not only fundamental for identifying problems, it is also fundamental for creating effective ways to face these problems with solidarity.

1.2 QUALITY OF LIFE AND ENVIRONMENTAL CHANGES

1.2.1 Quality of Life Concept

According to Seixas et al. (2010) the concept of Quality of life (QoL) refers to a complex system that involves three important tenets. All must be present when considering QoL for a society:

1. The first pillar: need for education, efficient collective transportation, employment, food, sanitation, healthcare, and salaries that sustain the needs of the individual and their family.
2. The second states the need for culture, leisure, friendship, intact family relationships, satisfactory relationships in the workplace, and connection to nature.
3. The third addresses ethical-political practices and access to information about political involvement and participation in collective causes and the local management of government and urban life (Seixas et al., 2010).

Based on these points, Seixas et al. (2010) and Romice et al. (2017) stated that it is possible to understand QoL as *the sum of the objective and subjective conditions in the everyday life of an individual.* Mitchell (2000) suggests six areas influenced by quality of life: health, safety, personal development, community development, natural resources, goods and services, and the physical environment.

Recognition of the relationship between QoL and environmental problems is a recent development in contemporary discussion (Seixas et al.,

2010; Félonneau & Causse, 2017; Fleury-Bahi et al., 2017). The human sciences and environmental sociology have only recognized this intimate relationship since the 1980s. This has most likely come about because of how environmental degradation has impacted the lives of populations in different regions of the planet.

The environment has become a central theme of important worldwide discussions, and the concept of QoL is relevant to this debate. When the concept of QoL was associated with the environment it became an approach to analyse modern society. It then began to attract attention on an international level. Initially the interest was focused primarily on the close relationship between QoL and health. Later, the strong interest in health issues was joined by a growing interest in urbanization and citizenship.

These conditions include the socio-environmental transformations that society undergoes on both a local and global level. In this daily struggle, the individual encounters information that either affirms or refutes these transformations. The individual can politically legitimize day-to-day life or alienate himself/herself. They can construct healthy or pathological conditions of life. The concept of QoL adopted here can contribute to a greater understanding of places and situations where socio-environmental transformations are evident – not only from the environmental point of view but also from the social, cultural, and political points of view. Thus, to Marchand et al. (2017) QoL is associated with health on one hand, and perceived environmental quality on the other.

The concept of QoL, according to Páramo (2017), takes on a special meaning when considering an urban environment. This is true, not only because urban environments account for the highest proportion of human settlements and the highest concentration of the world's population, but also because such population concentration creates social conditions that require planning and management to ensure the well-being of people and the future of civilization.

To Seixas et al. (2010) analysis of QoL contributes to the understanding of situations and places where problems are present. Social, cultural, political, and environmental elements must all be considered. We recognize, however, that although these elements are part of the project of modernity it hasn't been possible to achieve them for all society and they have been restricted to certain social segments. We especially find this in emerging countries. As a result, modernity and its problems, and the condition of the subjective in modern society, are fundamental issues that must be considered when thinking about QoL. This analysis shows unavoidable changes to everyday life by recognizing the degradation of the environment on a planetary scale. In addition, new territorial reconfigurations and transformations to culture impose countless objective and subjective daily challenges and difficulties for people. Bauman (2001) claims that it is unwise to deny or underestimate what he calls *liquid modernity*. It has produced, and continues to produce, profound changes to the human condition.

The link between QoL and the environment is becoming increasingly significant. At a local level for example, we have the effects that pollution and noise have on QoL. And, on a global level there is the central question of climate change with its harmful consequences for humans and the planet. The problems caused by pollution are extremely important. However, the lack of basic human needs such as food, water, shelter, and safety are of even greater concern because of their more immediate and drastic impact on QoL (Fleury-Bahi et al., 2017).

Quality of Life is defined as both a physical and a psychological state that provides human beings with the feeling of being satisfied with a given environment. It is generally considered to be the result of the interactions between several factors (health-related, social, economic, and environmental) which influence the development of individuals and society. The concept of QoL also allows the adoption of an interdisciplinary approach. The very nature of QoL places it at the junction of many disciplinary fields. This concept is used in many human and social science disciplines as well as in technological applications. These many fields have varying perspectives, and their definitions and meanings are not always the same. This situation creates ripe opportunity for collaboration.

To Pol et al. (2017) community life, social cohesion, empowerment, and participation are seen as prerequisites for the efficacy of any environmental intervention intended to improve citizens' Quality of Life. There is certainly not much quality possible in modern life without participation, citizenship, and agency. At the same time, it can be difficult to promote and sustain the participatory process if the potential participants are unable to perceive any possibility of improvement in their QoL.

1.2.2 *Quality of Life in a Climate Change Context*

Global temperatures, peak oil predictions, species extinctions, deforestation, glacier melting, desertification, and the increased concentration of carbon dioxide in the atmosphere, are already perceived and experienced as losses for many people. What further losses are coming to humanity that have precise timescales, associated calculations of harm, multiple projected scenarios of scarcity, and other related problems? There is a need for change and therefore action (Adams, 2016).

Global climate change (GCC) is a dynamic, multidimensional system of changes in environmental conditions that will likely influence human behaviour and quality of life (Evans, 2019). According to Obradovich et al. (2018) social, economic, and physical systems are critical determinants of psychological wellbeing. By disrupting these systems, climate change is likely to exacerbate known risk factors for mental disorders like anxiety, stress-related conditions, and mood disorders. Such psychological disorders reduce QoL, worsen overall health, and diminish productivity. Even subclinical levels of distress can impair psychological and immunological functioning and reduce the ability to cope with adversity.

Adapting to and coping with climate change is an ongoing and ever-changing process that involves many intrapsychic processes that influence reactions to, and preparations for, the adverse impacts of climate change. These include chronic events and disasters. Psychological processes include making sense of adverse climate change impacts, and appraising impacts and resources. They also include possible coping responses and emotional and motivational processes that are related to the need for security, stability, coherence, and control (American Psychological Association Task Force on the Interface between Psychology and Global Climate Change, 2010: 7).

To Obradovich et al. (2018) anthropogenic warming is increasingly presenting extreme meteorological conditions in any given year that effects people's mental health. Those who are more vulnerable, especially those with lower incomes, including women, may have less means to lessen the effects of adverse temperatures, and are therefore more susceptible to certain mental health difficulties.

The authors (Obradovich et al., 2018), also highlight that climate change is also likely to increase the rates of year-over-year and decade-over-decade warming of local climates and the chronic stressors that such warming produces. As a result, longer-term warming may have cumulative detrimental impacts on individual mental health. Additionally, climate change is also likely to amplify the frequency and intensity of acute climatic events, like tropical cyclones. Obradovich et al. (2018) conclude that climate change may amplify the society-wide mental health burden because of the acute environmental threats produced by the warming of natural systems. Given the vital role that sound mental health plays in personal, social, and economic wellbeing, as well as to the ability to address pressing personal and social challenges, their findings provide added evidence that climatic changes pose substantial risks to human systems and to the QoL.

Several studies (Obradovich et al., 2018; Seixas & Nunes, 2017) have recently indicated that the impacts of climate change are likely to effect and undermine mental health through a variety of direct and indirect mechanisms. Studies show that hotter temperatures and added precipitation both worsen mental health, and that multiyear warming is connected to an increased prevalence of mental health issues. In the same vein, exposure to tropical cyclones, which are likely to increase in frequency and intensity in the future, has been linked to worsened mental health. These results provide added large-scale evidence to the growing number of studies linking climate change and mental health.

Geographically, according to Albouy et al. (2016), climate effects the desirability of different locations and the quality of life offered there. Few seek to live in the freezing tundra or oppressively hot deserts. Given the undeniable influence climate has on economic decisions and welfare; valuing climate amenities not only helps us to understand how climate effects welfare and where people live, but also helps to inform policy responses to climate change. Global Climate Change (GCC) threatens to alter local climates, most obviously by raising temperatures. The impacts on wellbeing of higher temperatures are ambiguous. Households may suffer from hotter summers but benefit from milder winters. Ultimately, these impacts depend on where households are located, the changes in climate amenities they experience, and how much they value these changes, (Albouy et al., 2016). To Evans (2019), common psychological reactions to extreme weather and disasters, (such as prolonged drought, floods, and hurricanes or typhoons), include depression, posttraumatic stress disorder (PTSD), anxiety, and heightened family tension.

Current estimates suggest GCC will lead to approximately 150 million people being displaced worldwide in the next 50 years due to coastal flooding and soil degradation from drought and flood-related soil erosion. Evans (2019) mentions that destruction of habitat not only alters the structure and predictability of daily life, but also disrupts attachment to place. It is important to also recognize that disasters can severely strain socially supportive relationships. Common sequels to weather disasters are displacement from home and school, prolonged stress, and elevated household tension and conflict.

Another aspect to consider, according to Evans (2019), is rising air pollution that is the result of several aspects of GCC. Heavy precipitation increases molds and other bio-allergens; drought contributes to wildfires, which elevate fine particulates and gaseous pollutants; and higher temperatures accelerate the production of photochemical oxidants such as ozone. Research pertinent to the potential impacts of GCC on air pollution and human behaviour, has focused on awareness and concern for mental health in relation to outdoor recreation, (Evans, 2019). To Evans (2019), it is important to consider that changes in air quality are related to changes in the quality of life. Poor health and premature death linked to respiratory infections, heart disease, strokes, and lung cancer stem from specific forms of particulate air pollution (Adams, 2016).

Changes in climate have what is called a 'multiplier effect' leading to changes in other systems, such as social systems and soil systems. Gradual changes in quantity can suddenly tip over into changes of quality. One example of this is: when a human being who, after experiencing ongoing increased levels of stress, suddenly experiences a breakdown that creates an entirely new emotional state. Understanding the Earth as a dynamic system in which quantitative and qualitative changes interact, has informed the notion of "dangerous climate change". The belief that, with global average temperatures increasing more than two degrees, also comes the increased possibility of sudden and uncontrollable systemic ruptures (Hoggett, 2019).

There is a common perception that our response to date has been wholly insufficient. Inaction tends to be understood as resistance to a significant shift in consumption habits and the lack of a broader and sustained 'public response'. Public response includes social movement activity, behavioural changes, and public pressure on governments. If asked, almost everyone is in favour of protecting and preserving the environment and slowing climate change. However, people do not act consistently based on their declared values and best interests (Adams, 2016).

Human activity is both centrally implicated in causing climate change, and significantly affected by it. The Anthropocene also rebounds and reverberates back through human life. There are pervasive issues involved in uncritically homogenizing the role of 'humanity' in ecological crises. This is both in terms of responsibility and vulnerability (Adams, 2016). The author (Adams, 2016) also highlights that people's perceptions and understanding of climate change, do not necessarily correspond to scientific knowledge … because they are more vulnerable to emotions, values, views, and unreliable sources … Adams (2016) also mentions the following IPCC list of likely predictions:

1. Risk of death, injury, ill-health and disrupted livelihoods in low-lying coastal zones, small island developing states, and other small islands due to storm surges, coastal flooding, and rising sea levels.
2. Risks of severe ill-health and disrupted livelihoods for large urban populations because of inland flooding in some regions.
3. Systemic risks due to extreme weather events leading to the breakdown of infrastructure networks and critical services such as electricity, water supply, and health and emergency services.
4. Risk of mortality and morbidity during periods of extreme heat, particularly for vulnerable, urban populations and those working outdoors in urban or rural areas.
5. Risk of food insecurity and the breakdown of food systems linked to warming, drought, flooding, and precipitation extremes, particularly for poorer populations in urban and rural settings.
6. Risk of loss to rural livelihoods and income, due to insufficient access to drinking and irrigation water and reduced agricultural productivity, particularly for farmers and pastoralists with minimal capital in semi-arid regions.
7. Risk of loss of marine and coastal ecosystems, biodiversity, and the goods, functions, and services they provide for coastal livelihoods; especially for fishing communities in the tropics and the Arctic.
8. Risk of loss of terrestrial and inland water ecosystems, biodiversity, and the goods, functions, and services they provide for livelihoods.

The IPCC suggests mitigation that follows a simple enough logic: make changes to the sectors emitting the most greenhouse gases. However, these sectors overlap with a complex set of 'drivers' that include, according to IPCC, 'population size, economic activity, lifestyle, energy use, land-use patterns, technology, and climate policy' (IPCC, 2014a, 2014b). The IPCC (IPCC, 2014a) indexes various potential interventions targeting these drivers:

- large-scale changes to energy systems, (such as, improved efficiency, increased supply from renewables and carbon dioxide capture storage technology and changes to consumption patterns.
- land use (such as, afforestation and crop, livestock, and soil management).
- transport mode (such as, changes in transport mode, energy efficiency and vehicle performance improvements).
- buildings (such as, use of new technologies, changes in lifestyle, culture, and behaviour towards lower energy demand).
- industry (such as, energy efficiency, improvements in production, re-use, and recycling of materials).
- waste (such as, treatment technologies, recovery of energy from waste and efficient recycling).

Some of these changes are already underway and have achieved reductions in CO_2 emissions (IPCC, 2014b). However, the nature and extent of mitigation attempts required to be successful, (i.e., have a hope of keeping warming below 1.5 °C), over the remainder of this century are complex, contested, and uncertain. This is unsurprising considering the magnitude of the change required.

Another important point to address is social inequality and the living conditions of vulnerable populations. These populations are often found living in sensitive or degraded environmental areas where there is an absence of, or non-effective use of, environmental policy. We see this situation amongst indigenous, Quilombo, and Riverside peoples. Woodward et al. (2014) and Oxfam (2019) have emphasized that the most effective means of reducing vulnerability in the near-term are programs that implement basic public health measures such as:

- provisions for clean water and sanitation,
- secure essential health care services,
- increased capacity for disaster preparedness and response,
- actions to alleviate poverty.

Another important question is the adoption and implementation of effective environmental policy appropriate for regions vulnerable to environmental threats. Also important is the creation of action that stimulates and brings about the correct use of natural resources in these regions (World Economic Forum, 2020). Taken together these measures collaborate to maintain the quality of life of these populations.

References

Adams, M. (2016). *Ecological Crisis, Sustainability and the Psychosocial Subject – Beyond Behaviour Change*. Palgrave Macmillan.
Albouy, D., Graf, W., Kellogg, R., & Wolff, H. (2016). Climate Amenities, Climate Change, and American Quality of Life. *Journal of the Association of Environmental and Resource Economists (JAERE), 3*(1), 205–246. https://doi.org/10.1086/684573
American Psychological Association (APA). (2010). Psychology & Global Climate Change – Addressing a Multifaceted Phenomenon and Set of Challenges. *Report of the American Psychological Association – Task Force on the Interface Between Psychology and Global Climate Change*. Available: http://www.apa.org/science/about/publications/climate-change.aspx
Bauman, Z. (2001). *Modernidade Líquida* (pp. 23–63). ZAHAR.
Evans, G. W. (2019). Projected Behavioral Impacts of Global Climate Change. *Annual Review of Psychology, 70*, 449–474. https://doi.org/10.1146/annurev-psych-010418-103023
Félonneau, M. L., & Causse, E. (2017). Pro-environmentalism, Identity Dynamics and Environmental Quality of Life. In G. Fleury-Bahi, E. Pol, & O. Navarro (Eds.), *Handbook of Environmental Psychology and Quality of Life Research* (pp. 211–228). Springer Nature. https://doi.org/10.1007/978-3-319-31416-7_12
Fleury-Bahi, G., Pol, E., & Navarro, O. (2017). Introduction: Environmental Psychology and Quality of Life. In G. Fleury-Bahi, E. Pol, & O. Navarro (Eds.), *Handbook of Environmental Psychology and Quality of Life Research* (pp. 1–8). Springer Nature. https://doi.org/10.1007/978-3-319-31416-7_1
Hoggett, P. (2019). 1. Introduction. In P. Hoggett (Ed.), *Climate Psychology. On Indifference to Disaster, Studies in the Psychosocial*. https://doi.org/10.1007/978-3-030-11741-2_1
IPCC. (2014a). Summary for Policymakers. In C. B. Field, V. R. Barros, D. J. Dokken, K. J. Mach, M. D. Mastrandrea, T. E. Bilir, M. Chatterjee, K. L. Ebi, Y. O. Estrada, R. C. Genova, B. Girma, E. S. Kissel, A. N. Levy, S. MacCracken, P. R. Mastrandrea, & L. L. White (Eds.), *Climate Change 2014: Impacts, Adaptation, and Vulnerability. Part A: Global and Sectoral Aspects. Contribution of Working Group II to the Fifth Assessment Report of the Intergovernmental Panel on Climate Change* (pp. 1–32). Cambridge University Press.
IPCC. (2014b). Summary for Policymakers. In O. Edenhofer, R. Pichs-Madruga, Y. Sokona, E. Farahani, S. Kadner, K. Seyboth, A. Adler, I. Baum, S. Brunner, P. Eickemeier, B. Kriemann, J. Savolainen, S. Schlömer, C. von Stechow, T. Zwickel, & J. C. Minx (Eds.), *Climate Change 2014: Mitigation of Climate Change. Contribution of Working Group III to the Fifth Assessment Report of the Intergovernmental Panel on Climate Change* (pp. 1–30). Cambridge University Press.
Krenak, A. (2020). *Ideias para adiar o fim do mundo*. Companhia das Letras.

Marchand, D., Weiss, K., & Zouhri, B. (2017). Emerging Risks and Quality of Life: Towards New Dimensions of Well-Being? In G. Fleury-Bahi, E. Pol, & O. Navarro (Eds.), *Handbook of Environmental Psychology and Quality of Life Research* (pp. 531–542). Springer Nature. https://doi.org/10.1007/978-3-319-31416-7_29

Mitchell, G. (2000). Indicators as Tools to Guide Progress on the Sustainable Development Pathway. In R. J. Lawrence (Ed.), *Sustaining Human Settlement: A Challenge for the New Millennium* (pp. 55–104). Urban International Press.

Obradovich, N., Migliorini, R., Paulus, M. P., & Rahwan, I. (2018). Empirical Evidence of Mental Health Risks Posed by Climate Change. *PNAS, 115*(43), 10953–10958. https://doi.org/10.1073/pnas.1801528115

Oxfam GB. (2019). *Public Good or Private Wealth?* Oxfam Briefing Paper – January 2019. https://doi.org/10.21201/2019.3651

Páramo, P. (2017). The City as an Environment for Urban Experiences and the Learning of Cultural Practices. In G. Fleury-Bahi, E. Pol, & O. Navarro (Eds.), *Handbook of Environmental Psychology and Quality of Life Research* (pp. 275–290). Springer Nature. https://doi.org/10.1007/978-3-319-31416-7_15

Pol, E., Castrechini, A., & Carrus, G. (2017). Quality of Life and Sustainability: The End of Quality at Any Price. In G. Fleury-Bahi, E. Pol, & O. Navarro (Eds.), *Handbook of Environmental Psychology and Quality of Life Research* (pp. 11–39). Springer Nature. https://doi.org/10.1007/978-3-319-31416-7_2

Romice, O., Thwaites, K., Porta, S., Greaves, M., Barbour, G., & Pasino, P. (2017). Urban Design and Quality of Life. In G. Fleury-Bahi, E. Pol, & O. Navarro (Eds.), *Handbook of Environmental Psychology and Quality of Life Research* (pp. 241–273). Springer Nature. https://doi.org/10.1007/978-3-319-31416-7_14

Seixas, S. R. C., & Nunes, R. J. (2017). Subjectivity in A Context of Environmental Change: Opening New Dialogues in Mental Health Research. *Subjectivity, 10*, 294–312. https://doi.org/10.1057/s41286-017-0032-z

Seixas, S. R. C., Hoeffel, J. L. M., Bianchi, M., & Santos, A. (2010). Qualidade de vida, ambiente e subjetividade na APA Cantareira. In J. L. M. Hoeffel, A. A. B. Fadini, & S. R. C. Seixas (Orgs.) *Sustentabilidade, Qualidade de Vida e Identidade Local Olhares Sobre as APAs Cantareira, SP e Fernão Dias (MG)* (pp. 115–134). RiMa.

United Nations (UN). (1948). *Universal Declaration of Human Rights (UDHR)*. Available: un-org/en/about-us/universal-declaration-of-human-rights

United Nations (UN). (2015). *Transforming Our World: The Agenda 2030 for Sustainable Development*. Available: un.org/ga/search/view_doc.asp?symbol=A/RES/70/1&Lang=E

Woodward, A., Smith, K. R., Campbell-Lendrum, D., Chadee, D. D., Honda, Y., Liu, Q., Olwoch, J., Revich, B., Sauerborn, R., Chafe, Z., Confalonieri, U., & Haines, A. (2014). Climate Change and Health: On the Latest IPCC Report. *The Lancet, 383*, 1185–1189.

World Economic Forum – WEF. (2020). *Nature Risk Rising: Why the Crisis Engulfing Nature Matters for Business and the Economy*. World Economic Forum.

CHAPTER 2

The Importance of the Concept of Subjectivity: New Lines of Research

Abstract The central objective of this chapter is to analyse the role and importance of the concept of subjectivity and its close relationship to environmental and climate change. From the identification of a gap present in the literature, we seek from an interdisciplinary approach to bring a new perspective of analysis and also a contribution to the theme. The principals aspects in this chapter are: Approaches, Disciplinary dimensions: expanding views and concepts, Climate change and subjectivity, The construction of climate change narratives: 'nightmare' or 'false comfort', Climate change and psychoanalysis, The cultural narrative about climate change and subjectivity: Important Findings in Convergent Approaches, and Some important considerations about objective advances in the relationship of subjectivity, psychoanalysis, and climate change: and the construction of networks.

Keywords Subjectivity · Concept · The conceptual gap importance · Theory approach · Climate change

> In more and more spheres of human existence – politics, consumption, culture, and so on – there is occurring a displacement of the idea of the human subject as sovereign, and as subject to no normative claims which are not derived from what we, humans, want. In its place is emerging a recognition

that all human projects have to take account of natural limits, and of the intrinsic value of non-human nature. Once again, the world does indeed tell us what to do. (Szerszynski,[1] 2000)

2.1 Approaches

Even though we know that the history of consciousness and the recognition of environmental problems on a planetary scale, as well as the importance of building debate around this, was born in the contemporary world and began in the 1990s (UN, 1992) the warnings actually came much earlier, in the 1960s (UN, 1968). In the twenty-first century, these warnings increased and the question of environment was recognized on a global scale through innumerable reports and World Conventions about environmental degradation, atmospheric pollution, squandering of natural resources, health impacts, and quality of life for the general population (UNEP, 2002; UN, 2005).

This movement in global terms, has become even more intense since Rio 1992 (UN, 1992). In Rio de Janeiro, Brazil – twenty years after the first global conference on environment in Stockholm, Sweden in 1972 (UN, 1972) – Rio 92 convened. Rio 92 is considered a significant milestone for the planet in relation to the awareness of environmental questions. At the same time, the bringing together and use of diverse scientific disciplines, has brought significant advances to the understanding of the big questions that impact humanity and biodiversity. Today, with advances in science and communication, we have many ways to identify and communicate risks around the world. And, we are capable of identifying the large gaps that still exist in the twenty-first century.

One gap we identified is the relationship between environmental questions, environmental impacts, and environmental approaches to *subjectivity* and the human psyche. Although there have been important contributions from a number of places, after 2 years of research (Barbosa, 1996; Seixas et al., 2010; Seixas & Nunes, 2017), we have discovered several missing points necessary to establishing this important relationship. We also discovered that much still needs to be identified and clarified in order to establish the support needed for dealing with the great

---

[1] Szerszynski, B. (2000). On knowing what to do: Environmentalism and the modern problematic. In Lash, S., Szerszynski, B., & Wynne, B. *Risk, Environment & Modernity*. London: SAGE (pp. 104–137).

challenges which environmental changes impose on the planet and, in turn, the quality of human life.

Since the 1990s, many authors have called attention to the impacts of environmental change on human health (Marván & López-Vázquez, 2018; Liang et al., 2017; Cardwell & Elliott, 2013; Ebi, 2013; McMichael et al., 2003; Dutta & Chorsiya, 2013; Ermert et al., 2013). Some have alerted about the impacts on mental health as well, (Van Susteren, 2018, Clayton et al., 2015, Coverdale et al., 2018, Wei-Lun et al., 2018). We had the opportunity to draw attention to this in Seixas and Nunes (2017) but the identification and unfolding of these impacts on *subjectivity* or psyche is something we have only been able to identify recently. Now we have the opportunity to delve into these concepts here in this book.

In 2017 an article was published (Seixas & Nunes, 2017) that was the fruit of an extensive revision of scientific literature on this subject. This work underscored these gaps and the ways in which science and other disciplines are positioning themselves to resolve them. Current complementary theoretical and methodological proposals can collaborate and contribute to the scientific advancement of this question.

When we returned to the subject in this publication, it was clear that groups have emerged in the last five years that work with concepts akin to our vision of *subjectivity*. There are an increasing number of authors that explain theoretically and methodologically, the impact of the environment on the human subject, which goes beyond a singular relationship with nature, and speaks to the meaning of this relationship to life in general. We will discuss this in this chapter.

Seixas and Nunes (2017) tried different ways of approaching this theme and in some ways filled the gap. However, they were never able to effectively deal with the question because they didn't touch directly upon the *subjective* dimension of the human psyche and mental health. We will talk about *subjectivity* and explain definitively the direct relationship between sociology and psychoanalysis.

The important roadmap for this debate was established by a group of research scientists who are also psychoanalysts. They have an existential concern with the environment; the degradation of natural resources; and specifically with climate change and its impact on nature, the economy, human subjectivity, and the role that denial of these problems plays in society.

2.2 Disciplinary Dimensions: Expanding Views and Concepts

Reaffirming Seixas and Nunes (2017: 295), "*The concept of subjectivity plays a fundamental role in this article by offering a new basis from which to examine our understanding of mental health from an interdisciplinary perspective that goes beyond biology and climate science.*" In our opinion, what remains uncertain is the difference between the objective dimensions of claims about the mental health impacts of climate change, and *subjectivity* as it relates to social construction. Same is true for the recognition of suffering and the legitimacy of its associated factors – as understood from collective and individual responses.

The concept of *subjectivity* is a powerful example of our capacity to understand the relationship between environmental change and mental health. It goes beyond the limits of specific disciplines (Seixas & Nunes, 2017: 300). We suggest that in the final analysis this should be a new political project. A project where the production of *subjectivity* is the result of negotiated ideas. A committed project that embraces the fact that practices guiding sustainable development and resilience may be undesirable in the face of long-lasting injustice (Seixas & Nunes, 2017: 307).

There are other approaches that can bring important contributions to the conceptual understanding of *subjectivity*. Rey (2017) in his critique of psychological theory, shows that in different approaches to the concept of *subjectivity*, (such as psychoanalysis, Gestalt, and the post structuralist approach), there are many commonalities. He discusses in depth his specific proposition that it is from a historical, cultural, and philosophical viewpoint that the concept of *subjectivity* began to be ontologically understood.

The author makes the criticism that in the majority of these theories *subjectivity* has been used in reference to specific processes and phenomena, indicating that advances haven't been made in the construction of a general theory of *subjectivity*. The author also emphasizes that *subjectivity* has been treated within a Cartesian tradition allied to individualism and connected to the tradition of critical psychology. He believes this helped in the rejection of the concept. In this way, these critical theories have left out the heuristic value of *subjectivity* in the study of processes that can't be exhausted by discourse or language (Rey, 2017: 1).

The author proposes the emergence of a new approach to *subjectivity* based on the historical tradition of Psychology. From this perspective, *subjectivity* is defined by units of symbolic emotions that are generated in every human being. Based on this definition, the author states that institutionalized orders can be subverted by productions that represent new social paths, and proposes that *subjectivity* should be defined as a human production capable of transcending the apparent limits of objective existence (Rey, 2017: 2).

In another seminal work, Rey (2019) adds an important perspective to the ways in which the use of *subjectivity* has been impoverished in psychology, as well as in other disciplines. The author emphasizes that recognition of the symbolic character of human phenomena occurred relatively late in relation to psychology, linguistics and anthropology. Early on, its recognition was so radical it led psychologists to deny the majority of concepts traditionally used in psychology.

Drawing upon theoretical traditions that have advanced the understanding that the human psyche is a historically and culturally engendered phenomena; the author proposes a new definition of *subjectivity* as a phenomena that emerges as a result of symbolic forms that are socially-historically situated, and out of which emerge concepts such as discourse, deconstruction, and systems of dialogue-communication. This orients subjectivity to descriptions of human processes that aren't exhausted by these concepts, and where the study of human realities is complimented by approaches that are more ample and complex (Rey, 2019). In this way, Rey proposes that the concept of subjectivity within a historical-cultural approach – far from opposing the concept of discourse – instead compliments it by advancing a new theoretical system capable of generating new knowledge and practices related to specific human aspects of social or individual phenomena (Rey, 2019: 179).

The author radically implies that subjectivity mixes with discourse and forms a new system in which the epicentre is the forms, processes, and symbolic realities that characterize human existence. The human cultural world is formed and developed through symbolic processes and realities that are inter-related by a cosmos of symbolic constructions, discourses, and social representations. For example, the universe of concepts, myths, and preconceived ideas that characterize human existence (…) making the discourse the absolute ontological definition of human phenomena (Rey, 2019: 179).

Since gaining understanding by reading the works of Rey (2017, 2019), it has been interesting to observe complimentary understandings of subjectivity, its role and interdisciplinary importance, and the unveiling of disciplines that best contribute to – or should contribute to – understanding it. Chancer and Andrews (2014) in their work – The Unhappy Divorce of Sociology and Psychoanalysis – show that sociology and social sciences in general have a lot to gain by incorporating, rather than excluding, psychoanalysis and psycho social approaches in a diversity of social topics – specifically related to the social problems of inequality, poverty and climate change.

The authors also affirm that the psyche and psychoanalytic territory is perceived as an area that exclusively, and with priority, treats the individualized and the individual. They conclude that, in order to approach the subjective in its sociological dimension, psychoanalytic methods shouldn't be considered, as they have nothing in common with social research. This tradition of thought in the second half of the twenty-first century has caused many social researchers to hesitate to employ or encourage the use of psychoanalytic tools and concepts in sociological research. Notwithstanding, according to Chancer and Andrews (2014: 1) this often impoverishes a more creative approach to the study of subjectivity. In other words, this hesitation creates an imposition that limits and makes more difficult, the carrying out of the research needed to find answers to our current issues. This is characterized by anxiety, exacerbated consumption, difficulty in recognizing human rights, conservative values, and difficulty in finding academic work that offers positive answers to human quality of life (Chancer & Andrews, 2014: 3).

2.3 Climate Change and Subjectivity

In this section we will talk about climate change as a result of human behaviour. With this assertion now as consumed fact, and in face of the terrible difficulties our planet is experiencing, it is very important to discuss climate change, (the most important environmental change), and its relationship with human subjectivity.

In 2013, Dunlap alerted that since 2000 scientists have brought attention to global warming, including testimony given to the United States Senate (McCright & Dunlap, 2000; Dunlap, 2013). Unfortunately, not only was little progress made in dealing with global warming but in the years that followed the problem worsened. Greenhouse gases increased, which generated more heat and brought increasingly negative risks to social and natural systems (Dunlap, 2013).

Since 2007, however, science has confirmed what used to be a hypothesis (Fritze et al., 2008; IPCC, 2007; Giddens, 2009; Dunlap, 2013; Randall, 2005; Few, 2007; Ebi & Semenza, 2008). Randall (2005) had already warned that, with the exception of a few sporadic contrary positions, scientific agreement had been reached about the priority and seriousness of climate change and the increasing anxiety about environmental degradation and its connection to world poverty. (Randall, 2005: 165–166). The author points out that research proves that 60% of the services given by the world's ecosystems, and the processes and products from nature that sustain life, are being used unsustainably (Randall, 2005: 166). The steps necessary to mitigate those effects, principally in relation to the reduction of carbon emissions, are not being effectively taken. It will require great changes in the patterns of economic activity, lifestyle, and behaviour of people in developed countries for effective changes to be made. The author affirms that if no action is taken, the effects on natural and human systems will be catastrophic and irreversible. Although this situation will not affect our generation directly, it will have a dire effect on future generations (Randall, 2005: 166). This assertion was made two decades ago.

Dunlap emphasized in 2013, that the complex nature of anthropogenic global warming (AGW), or in other words global warming caused by human beings, generates uncertainties in regards to the risks it represents. This makes it difficult for lay people to understand its causes, perceive its impacts, and take the actions necessary to help alleviate future heating. Rustin (2013) mentions that climate change "*has now become a biophysical-social phenomena because human beings have acquired the power to influence the climate through their activities*" (Rustin, 2013: 3).

Science has played a fundamental role in establishing the fact that global warming is occurring and that human activities do contribute to this warming. Science has also established that current and future events portend negative impacts on ecological systems. Despite these warnings, a significant number of society's population remain ambivalent or unconcerned. Many are in a position of denial regarding the need to taking effective measures to reduce carbon emissions, or promote changes in behaviour in regards to the use of natural resources and consumption patterns (Dunlap, 2013: 692).

2.4 The Construction of Climate Change Narratives: 'Nightmare' or 'False Comfort'

Randall (2005) has pointed out that, although science/scientists acknowledge the existence of climate change, the general population has not given this issue the importance it deserves. For the most part, people carry on as usual and don't give climate change much thought. "If the news about climate change and environmental degradation upsets people or causes them anxiety, they aren't showing it" (Randall, 2005: 166).

Our research on Brazilian consciousness has led us to mistrust the expression of anxiety and depression; and the construction of denial or silence, as affirmed by Randall. Research we have carried out (Seixas, 2008; Seixas et al., 2010, 2012, 2014, 2016; Seixas & Nunes, 2017), shows a tell-tale picture of unnamed symptoms in the population as a whole that are causing people to seek help from public and private health services. Antidepressants and anti-anxiety medications are being increasingly consumed by the Brazilian population. Perhaps this upset in public mental health is being attributed to personal, every day stressors, without considering that collective population issues could be the cause – for instance, a possible catastrophe that humanity is about to face. This position may be more connected to the population's denial over collective causes, leading to difficulty in recognizing global problems, which again leads to neglect in organizing ourselves to face these problems. It then follows that political action necessary to face this planetary political, social, and environmental moment will also fall by the wayside.

In the vision of Randall, it's interesting that despite widespread knowledge and a general concern with climate change, there remains a disconnect that doesn't allow individuals and society to see the urgent need for behaviour changes to avert and/or prepare for the impact of climate change. The author used one very interesting example of this in her article defending research done in 2004 in the United Kingdom by the BBC. This research showed that although the majority of British citizens had agreed that human activity is responsible for climate change, 43% don't believe they will be personally affected by it, and only 37% would agree to pay more for gasoline in order to mitigate the conditions fossils fuels impose on the planet (Randall, 2005: 167). This poses a great dilemma: In what way can science and scientists contribute to minimizing the impact of climate change and also bring awareness to society about the gravity of the situation in which we are living? What information can studies about subjectivity and psychoanalysis offer to a solution? This is the important discussion to which we wish to contribute.

It is impossible to deny that information has been manipulated in the construction of the narrative about climate change. Information offered to the public has been built around the demands of an economic system exclusively based on growth and the self-perpetuation of elites. These points are central to any discussion about environmental problems. Their extreme complexity and their existence make them vital to any and all debate on this issue.

Randall also called attention to the fact that, although there is much debate, few solutions have universally been agreed upon. It is likely that a great many people feel helpless and unprepared to deal with this situation and its effects (Randall, 2005: 167). In order to find solutions for this problem, we are led to the realm of psychoanalysis. Psychoanalysis can make a powerful contribution in helping to resolve the grave collective social problems that humanity has faced and will continue to face.

Segal (1987) has become an important reference when focusing on the grave social impacts humanity is experiencing. In his text, the psychoanalyst effectively examines the impact of the arms race on the collective psyche, and the role psychoanalytic reflection can play in this examination. His work has shed light that goes far beyond this specific question, which allows other authors to follow his lead in applying his conclusions to similar issues. Authors can approach other social issues with the same robustness that the subject of mass collective nuclear destruction promoted. After all, in terms of the extent of devastation, climate change is no less significant (Segal, 1987).

Rustin (2013) calls attention to two important topics that affirm our previous conclusions. First, he recognizes as being both real and exceptionally grave, that the risks from climate change are mostly consequences of human activity. Secondly, although it is clear that there are various disciplines that greatly contribute to the understanding of these risks, psychoanalytical thought can offer a different, behavioural understanding in the search for solutions. Rustin emphasizes that updating the psychoanalyst perspective in regard to the understanding of climate change, is far from being ignored or rejected. Over the past 3 decades, global warming has become a broadly debated issue in many arenas. It is an area of study that currently presents broad scientific consensus on the risks and impacts. This consensus deems it necessary for government, corporations, social movements, and citizens to assume responsibility, even if insufficient, for taking measures to limit and avoid the dangers of global warming. The author affirms that a psychoanalytic perspective can contribute to uncovering the unconscious anxieties and defence mechanisms underlying wide spread problems that indistinctly effect the planet (Rustin, 2013). Many concerns that have been openly expressed allow little room to affirm that the principal motivations and anxieties about these issues remain unconscious and in wait of recognition through psychoanalytic insight.

In his text, Rustin (2013), makes evident that one way to contribute to broadly understanding the relationship between human society and the natural world, is understanding that the present threat of climate change is the result of humanity's inherent destructiveness in relation to nature (Rustin, 2013: 174). Also, climate change is a threat on a much greater, global scale than more specific threats of the past. More significantly this means that the social, technical changes necessary to face climate change are equal to the greatest past industrial revolutions and will take decades to evolve and function to their capacity. We should also mention that the majority of new technologies may bring pathological side effects. The author affirms that the problem with climate change is that its damage tends to be immense and calamitous to the entire planet (Rustin, 2013).

The author also states, that even though 'technical' solutions for addressing climate change are being developed, viable solutions for environmental problems have not been addressed with the same force. These are problems that go beyond material reality, as in the case of social and political issues. Because these problems are not being focused upon with as much vigour, we can expect the worst outcomes Rustin (2013). When considering the general understanding of climate change and its influence on the future of an ever more global society and, despite the fact that concern regarding the wellbeing of all beings exists, there is little that leads us to believe necessary actions will be taken to face this crisis.

The different forms of action and the different ways of looking at the debate, take us back to another warning given by Randall (2009) that is directly related to a psychoanalytic approach that calls attention to the discourse on climate change. It states there are two parallel narratives – one about the climate change problems themselves; and another that is about possible solutions.

The author points out that the narrative about climate change problems, and the dramatic losses it can cause, are perceived as being in the future, or in places far removed from the western world. The author also points out that in writings that discuss solutions to the problems, these losses are not mentioned at all. She suggests that the division into parallel narratives is the result of a defensive process of division and projection that protects the public from its need to face and truly mourn the losses associated with climate change. In other words, it allows the public to remain in state of denial about the seriousness of the issue. This false sense of security produces images projected into the future that are accompanied by ineffective proposals for changes in the present (Randall, 2009:118).

The author, a psychoanalyst by training, suggests that a more sophisticated understanding of grief and loss processes will reframe the public narrative causing the liberation of energy needed for real and permanent change. Based on models of psychoanalysis for grief and loss began an empirical investigation using small groups in Cambridge, the United Kingdom. The goal of the investigation was to ascertain ways of building leadership and support for practical programs that would help the public accept changes in the environment. Such programs would build aspiration, culture, security and identity in order to face global problems of this magnitude (Randall, 2009: 118).

Randall points out the writings that broadly ignore the question of loss and, even though these narratives imply that if we don't take immediate action there will be catastrophic losses, they don't take into consideration – at least the possibility – that we are already suffering losses. The actions presently being taken to avoid catastrophic loss, need to consider the losses we are already suffering. Psychologically, we can understand this approach as a division and a projection. Psychoanalytical theory argues that the child gradually perceives that the 'adored mother' that attends to their emotional and physiological needs is at the same time the 'hated mother' that fails the child in the meeting of these needs. The ability to understand this, and the ability to deal with this ambivalence, is considered an important step towards mental health (Randall, 2009:119).

This concept can be applied to the planning of environmental protections in relation to future generations. Protections that guarantee a model for sustainable development. The difficulty of projections for the future perhaps lies in the need to guarantee the present generation Quality of Life, in order for it to be able offer support to future generations.

Randall reminds us that according to the psychoanalytic construction of subjectivity, disagreeable parts of the self, (such as undesirable knowledge and desires), can be separated and projected onto other people, times and places. The consequence of this is the construction of a world of extremes where good is idealized and bad becomes terrifying. Subjectivity then becomes a process where the fear of loss results in a division and projection onto the future, and the present continues to represent security, even though the costs for the future are terrifying. This process illustrates two dimensions: On one side, a nightmare and on the other side a false comfort that results in the creation of a powerful system of psychological defence (Randall, 2009: 119).

The author suggests that, in comparison to 'apocalyptic' representations of the future, narratives about solutions can be considered anticlimactic. The following neither alter patterns of economic growth, nor do they alter the present paradigm of economic development:

1. *Small steps* where everyone does their part in making changes in their relationship to nature, their use of natural resources, and their use of technology available for saving energy.
2. Transformation of large and small habits of consumption to natural and sustainable habits. Substituting electrodomestics for energy saving models, use of solar panels, and generally making action on climate change an integral part of the daily life of the consumer.
3. Developing faith in new technologies such as renewables, nuclear energy and geoengineering.
4. Decarbonization.
5. *The Happiness Tale:* Life will change, but society will be happy with the changes because with low consumption of carbon throughout the world we will have a greater sense of community, there will be more meaningful time for work, and we will have more quality time to spend with family (It is important to note that these steps or strategies will all rely heavily on technology). Economic growth won't be affected because these steps only require a change from carbon-based technology to a lower carbon technology. The lives of individuals will suffer little change (Randall, 2009: 119–120).

Can we really be sure that the best strategies are those that don't question the present economic model and maintain a powerful connection to technology? As long as the approach to sustainability:

- fails to recognize the importance of the principals of Agenda 2030;
- fails to deeply question capitalism which profoundly increases social inequality;

- exploits resources;
- and neither maintains nor constitutes the dignity of life,

This will be our challenge.

2.5 Climate Change and Psychoanalysis

An important aspect in this narrative is the intrinsic relationship between the psychological and social. The subjective dimension exposes this relationship. Weintrobe brings attention to the construction Randall calls, "*the third position, a situation in which placing psychological and social realities side-by-side allows for the emergence of points of creative integration*" (Weintrobe, 2013: 3). The author's premise suggests going beyond a vision centred on aspects of individual analysis on one hand, and social analysis on the other. Going with a premise that is constructed by different disciplinary approaches can be difficult, if not impossible to sustain. Consolidating two approaches where one approach has more privileges and analytic strength than the other – and will not necessarily join together at some point – make this approach non-viable. This is the aspect that Seixas and Nunes (2017) aimed to critique. Along this line, Weintrobe (2013) defends and emphasizes three analytical contributions that can explain this important position. These contributions reinforce the approach proposed in the construction of a 'third position' and which he defended in the work he organized:

1. Whether social or individual, all of the present approaches that humans have constructed are about their relationships, and conflict is a fundamental part of them.
2. The role of scientists' engagement in the problem of climate change is within the broader context of their involvement with questions of social justice. When human beings are the object of study, whether individual or as part of humanity, there are questions of moral conflict involving rights, responsibilities, and justice in relationship to oneself and to others.
3. Most important of all, is a common underlying drive that involves the search for a deeper meaning, postulating underlying structures that aren't manifested on a superficial level but give due weight to human complexity (Weintrobe, 2013).

It is important to emphasize that many of the environmental problems that involve the planet at this time, have a lot to do with humanity's anthropocentric view of itself. Weintrobe states that this has resulted in the philosophical vision that underlies the human belief that it is our right to have dominion over all other forms of life. This viewpoint powerfully legitimizes humanity's exploitive side as instrumental to our relationships. This includes our relationships with nature and animals. The fight for universal human rights is rooted in the fight that some human beings (such as the poor, women, and children) are inherently less equal than those authorized to dominate and exploit the 'not authorized'. This instrumental point of view can also affect how we take advantage of science and technology for our interests, and maintain those advantages for a privileged few and not for the collective as a whole (Weintrobe, 2013).

Lertzman (2015: XV) seeks to analyse how people go about reformulating the ways in which they are accustomed to thinking, and how they do this with the most serious ecological threats they face as a species and a planet. The author observes two lines of thinking: On one hand, a total absence of care or concern; and on the other hand, an unbearable level of anxiety and ambivalence that can frustrate the capacity for action and organization needed to effectively face these problems. From our perspective, this last aspect strongly illustrates the impact of the subjectivity of the population. This anxiety doesn't have a name, nor can it be diagnosed (as psychiatrists would like).

The author affirms that despite having worked as a professional in environmental communication in different social sectors (governmental, public, and private) she is always asking herself how to mobilize the public in a way that includes effective environmental actions. Interestingly, she points out that even in different spaces, the same underlying supposition persists – people don't care enough to act and the dilemma of how to motivate people to care persists. Up to this moment, however, no sector has produced an effective method that changes human behaviour in relation to conservation, protection, and restoration of the environment (Lertzman, 2015: 3).

Hoggett (2019: 3) points out that if we do nothing to maintain the COP 21 – Paris 2015[2] – accord there will be an average increase in

---

[2] The twenty first Climate Conference, or simply COP21, Paris 2015, was one of the largest and most important meetings ever organized by the United Nations, to discuss and determine global actions to combat climate change from 2020, replacing the Kyoto Protocol.

temperature of at least 3.0% by 2100. This combined with nationalism and the collapse of already timid international cooperation in relation to climate change, it is certain that projections will get worse. A great deal of the planet will become uninhabitable and unexpected negative impacts on our oceans and on agriculture will make the present massive migration nothing in comparison to what we will see in the future. This, without a doubt, will be a type of social, economic, and environmental collapse without precedence for humanity.[3]

Besides environmental changes, especially climate change and its negative impacts on human subjectivity, in December 2019 we saw Sars-COV-2 appear. In less than 3 months, it left its specific niche in Wuhan, China and spread all over the planet revealing our limitations in dealing with an event of such health, economic, social, and political magnitude. This pandemic gives humanity a rehearsal for the great catastrophes the planet could face with climate change.

It's important to note the fact that scientists who collaborated on the IPCC 2007 report had already advised of increased threats to human health directly related to climate change. There is strong evidence that exposure to climate change effects the health of millions of people, especially those with a low capacity to adapt. In other words, the poorest and other groups that are more vulnerable due to ongoing changes in the spacial distribution of certain vectors of infectious disease (IPCC, 2007: 10–11). Seixas and Ferreira (2020) pointed out that scientists had already given a crucial warning, (14 years ago), that is directly related to our present challenges. Viruses pathogenic to humans (ex. Ebola, H5N1 influenza, bird flu, etc.) occur naturally on the planet and live within a natural cycle in wild animals that host them. It's when human actions alter this cycle, that the virus suffers mutations and genetic recombination. Organisms that are not adapted to it are exposed. This is one of the exact hypotheses for what is happening now with the Corona virus in relation to humans and the unleashing of this pandemic (Seixas & Ferreira, 2020).

---

[3] It is important to note that in 2020, in China, COP 26 would be held, but due to the Covid-19 Pandemic it was postponed and will take place between October 31 and November 12, 2021, based in Glasgow, UK. Further details on the schedule see: https://ukcop26.org/. For details about the agenda and suggestions that will be discussed, see: https://2nsbq1gn1rl23zol93eyrccj-wpengine.netdna-ssl.com/wp-content/uploads/2021/06/COP26-Explained_.pdf

2.5.1 The Cultural Narrative About Climate Change and Subjectivity: Important Findings in Convergent Approaches

When we look at climate change and subjectivity, we must include a critique of capitalism. In other words, analysing climate change must take into account a critical position in regards to the hegemony of the capitalist model. That is the basis of all of our values and social behaviours, including our subjective postures.

In his article Finley (2019), offers an important critical analysis of the ideological position of decreased growth from a social ecological perspective. He affirms what Giorgios Kallis pointed out, highlighting the need to abolish the growth imperative that capitalism embodies today, and at the same time offer a critique of the conceptual foundations of this notion. The radical position the author takes, goes farther by saying that growth as an ideology reproduces a binary conception of society and nature. This is a position of opposition that is a key concept to growth as a political agenda, and one that is totally prone to appropriation for authoritarian purposes. As pointed out by various references recently highlighted, the author defends the imperative of considering the need for a social ecological position.

Euler (2019) is an excellent example of the above. In a theoretical article, the author begins by reconstructing an argument of critical value and affirms that capitalism is a form of society that is structurally unsustainable. The core of capitalism requires production of ever-increasing value, (commodity form, competition, profit maximization, private production). Based on this, the article calls for a fundamental social transformation and suggests innovative social practices (commons) with the potential to replace the commodity form as a social foundation. *Commons (social practices) are based on volunteering, autonomy, and satisfying needs.* Employing this model of commons eliminates the built-in growth compulsion of capitalism. Therefore, the article concludes, the commons may allow humanity to deal with the question of sustainability by creating social structures that include the possibility of a solution.

Hughes (2010) states that since 1975 there has been agreement amongst climate scientists that global warming is caused mostly by human activities and attitudes. The shift in thought needed to mitigate global warming – such as reduction of greenhouse gases – is a primary subject of national and world agendas. We see this in today's world climate conferences. (See notes 4 and 5) It is important to emphasize that, because of the sophistication of computational modelling – a result of scientific advances in the last decades – scientists are now able to confirm that unprecedented global warming, caused by an increase in carbon dioxide, is occurring on the planet. The suggestion that if these alterations continue, they will create negative impacts from increased temperature; sea level rise that results in flooding of islands and coastal areas; changes in precipitation; interruption in the flux of fresh water, caused by the loss of ice; stress on agriculture, forests, wild animals, coral reefs and fish, coming from increased acidity and rising ocean temperatures.

The author cautions that, although the United Nations created the Intergovernmental Panel on Climate Change (IPCC) in 1988, it has been quite conservative in its structure and cautious about the aspect of interfering in human activity, at least up until 2007. The international discussions in Rio (1992), Kyoto (1998), and Copenhagen (2009) indicated the importance of collecting current environmental information. They also illustrated the need for action aiming to combat global warming. However, political pressure dominated the negotiations and most nations were swayed into taking action that did not place significant importance on the global warming crisis. Nations commitments to accompany the necessary actions even where there were accords has been neither real or clear (Hughes, 2010). It is interesting to note that the Paris Conference (COP 21), tried to change this and promoted important agreements to improve planetary climate conditions. However, with the ascent of 'conservative' governments in the USA and Brazil after COP 21, the questioning of scientific proof regarding the role of human behaviour in continuing climate change became a form of political resistance. The existence of climate change/global warming was refuted by the presidents of these countries with the full purpose of not promoting the necessary attention to the measures that had been agreed on in 2015. Since 2021, with the ascension of the Democrat Joe Biden to the presidency of the USA, there was an immediate return to the Paris Accord and climate public policy became part of the agenda of that country once again. The existence of climate change/global warming was refuted by the presidents of these countries with the full purpose of not promoting the necessary attention to measures that had been agreed upon in 2015.

As the awareness of climate change probability has spread throughout the general public and economic measures to reduce carbon emissions have gained support, opposition to efforts to deter or mitigate global warming have emerged on at least three fronts:

1. Some scientists have pointed out glitches with the evidence and the models used for making the previsions. This is to be expected as critique is part of scientific investigation. Good scientific critique is positive because it leads to open discussion and additional investigation and testing of hypotheses. In the case of global warming, *the weight of scientific opinion has concluded that global warming is happening* and at least a great part of it is due to human activities.
2. Industry has a large share of the responsibility in human induced climate change and will have to absorb the costs and the enormous efforts needed to combat it. This includes fossil fuel industries, as well as car manufacturers. These industries have created advertising that debunks the idea of global warming. They have engaged in biased campaigns destined to sow doubt about the reality of human involvement in the creation of global warming. Most recently, they conveyed negative information about IPCC researchers in order to undermine the work of these scientists.
3. Right wing political organizations also actively create programs that oppose the mitigation of global warming. Such organizations fear that a reduction of greenhouse gas emissions will instigate regulation on both a national and an international level and these groups combat government intervention on principal. They see the questioning of climate change, and humanity's role in causing it, as a way of resisting increased government control.

The author poses the important question of who will most intensely suffer the negative effects of global warning. It won't be the wealthy because they have the resources to build defences or get out of the way. It's will be the common people, the workers and the Earth's most vulnerable who will encounter the most extreme disasters. Those who have no escape. Cities and rich nations can build dykes to hold back the rising seas, but the people of Bangladesh and the Maldives will have the choice to either flee or drown. This was clearly evident during and after Hurricane Katrina in New Orleans, United States, August 2005, when the poorest neighbourhoods suffered the most consequences.

The history of recognition of global warming and its implications on human society, is illustrative of the interaction between the growth of science and knowledge, and the competing interests of politics and economic entities. Science can evaluate potential measures for impeding negative changes, or at least lesson their magnitude. It can also evaluate courses of action that help human society deal with the negative effects of these probable changes. The decisions, however, on what measures will be implemented by government or business is historically linked to patterns of action directly related to their own short-term interests. Those sectors have notoriously turned a blind eye to the long term common good of humanity and the earth.

Correia (2016), nonetheless, pointed out something curious at COP 21 in Paris 2015, that the author described as 'climate revenge': Since Kyoto and the advent of limiting greenhouse gas emissions, the levels of greenhouse gases have actually *risen* from 360 (parts per million) to more than 400 ppm. This is despite the capital of some CEOs and the work of economists that, in the opinion of the author, have been involved in ongoing climate claims through the emergence of the international carbon market. The author believes the Paris Accord to have been a climate victory. Also, a complete victory for a market-based approach to climate change. Interesting possible scenarios:

- One where nothing happens
- Another where a radical climate movement emerges out of Paris that interrupts the market logic favoured by UN and corporate interests, and constitutes a real state of emergency.

Another important contribution to this topic are approaches that consider the human dimensions of the climate debate. This debate mostly comes from scientists in the humanities who are focused on social and political science. Such contribution can be observed in a special issue in 2010 by Szerszynski and Urry, in Theory, Culture & Society. Their set of 14 articles highlight this approach.

Both Szerszynski and Urry underscore the value of the human sciences to the climate change debate and the important contributions social scientists can make to this subject. Their special issue was published just after the emergence of three important documents for world debate. These documents serve as a background for this report and are approached in different manners by the 14 articles that make up the special compilation. The authors mention the important works of Lovelock (2006), Giddens (2009) and the IPCC report published in 2007.

Szerszynski and Urry (2010) point out the powerful impact of James Lovelock's book (2006). Based on his Gaia Hypothesis, the author maintains that for thousands of years human beings have exploited the planet with no concern for the impacts of their actions on the planet. Gaia, (earth as a living organism), in the hypothesis of the author, is hitting back and we have arrived at a point of no return (this was in 2006). A point where sustainable development is no longer possible *"despite all of our efforts to retreat in a sustainable form, we may not be able to impede global decline in a chaotic world governed by the lords of brutal war on a devastated Earth ..."* The authors mention Giddens (2009), the world-renowned social scientist who broadened this debate and gave social scientists and their contributions, the recognition they deserved. When acknowledging the importance of Giddens, it is worthwhile remembering and revisiting his work where he clearly states that *"Global Climate Change is one of the pillars of the 21st century agenda. (...) and that changes on course may have catastrophic consequences for the planet."* According to the author, the difficulty humanity has in creating effective measures is because climate change should be looked at and treated as a political question, and that all decisions must observe the economic and geopolitical context of the world (Giddens, 2009). Lastly, the authors highlight the role of the IPCC reports, with special emphasis on 2007. The report from 2007 was intended to be assertive and definitive in regards to the anthropogenic role in climate change. Before 2007 there was much reticence in this regard.

Based on the above references, Szerszynski and Urry (2010) explain how this special issue was organized. The reading of these three documents allows social theory to enter the debate of how land occupation and use, as well as the present economic model, generate high levels of 'greenhouse gases' that appear to be elevating the earth's temperature and transforming the future of humanity's way of life. Despite attempts to monitor and 'mitigate,' climate change will likely bring about drastic changes in social organization: the creation of a global assembly for international agreements, carbon atoms, markets, technologies, climate events, and social practices (Szerszynski & Urry, 2010: 1).

From the viewpoint of the authors, there is no doubt that the twenty-first century will be recognized and defined by the fundamental debate on climate, resources, and energy. Based on this, the articles were organized around three narratives present in the climate change debate (Szerszynski & Urry, 2010):

1. *Scepticism* that has challenged all the data accumulated by science on climate change. The sceptics suggest that climate changes could very well be due to 'natural' processes, like fluctuations in solar

activity, and not from 'anthropogenic' processes. They also claim that the climate change argument is motivated by science and the press. Others contest the enormous costs involved in any attempt to resolve climate change in comparison to other global challenges (Szerszynski & Urry, 2010: 1).

2. *Gradualism* involves allegations that the climate is changing but that these changes are relatively slow. Economies can adjust to reduce and adapt to changes by means of operational investments, and through insurance risk calculations that will induce individuals and societies to change their behaviour by means of appropriate incentives (Szerszynski & Urry, 2010: 1–2).

3. *Catastrophism* criticizes these two positions based on historical and archaeological data. It argues that the IPCC reports neglect the possibility of unpredictable extremes and climate changes that result in positive feedback loops. Catastrophism also presumes that many of these changes are already occurring and little can be done to impede various impacts that are already *in the system* (Szerszynski & Urry, 2010: 2).

Through its articles, this special edition sought to question these three narratives that more and more dominate global agendas and which, in our opinion, have been around for about a decade. They examine the rapid transformation of scientific understanding of *environmental danger*. They consider contributions made from the physical sciences in establishing *the unmistakable* truth of climate change and its establishment as the primordial environmental problem of this century. They also highlight the unique *power of science* to organize and mobilize the world to action and to create events focused on the crisis of the world's perception of global climate change. IPCC actions, organized by thousands of scientists and policy makers around the world, transformed the public and political debate despite facing great opposition, particularly from the USA, and especially since 2005. Recently, for example, the Pentagon suggested that climate change would result in global catastrophe that will cost millions of lives and that its threat eclipses global terrorism (Szerszynski & Urry, 2010: 2).

In this work we will prioritize certain analytic aspects from 7 of the special edition articles that are more directly related to the objectives of this book. Clark (2010) defends that climate change happens suddenly, abruptly and in an uncontrolled fashion. This claim has been defended by a great many scientists, principally by scientists related to the IPCC, who suggest that the evidence presents enormous challenges for the international negotiating and regulatory organizations involved in climate change. Up until now they have concentrated on gradual changes. The author states that one way of dealing with catastrophic climate change is to emphasize the volatility and unpredictability inherent in terrestrial processes, and the no less vulnerability and volatility inherent in the human body. In this manner, Clark will prioritize the question of environmental justice in a broader sense of what climate change means to human beings that have already suffered and have been victims of past episodes of abrupt climate change. Clark suggests that we experiment with broadening the idea of generosity amongst human beings.

Beck, who was an important Polish sociologist in the second half of the 1980s, gifted us with the construction of important concepts for thinking about contemporary society. The most important of these concepts is what he calls the risk society and the meaning of reflexive modernization.[4] In his text, Beck raised questions that are still relevant today (2021) and sought answers to these questions in 8 theses.

The author poses the following question "*why is climate change and ecological crisis, one of the most pressing questions of our times, not embraced with the same energy, enthusiasm, optimism, ideas, democratic spirit and forward thinking that tragedies of the past like poverty, tyranny and war have been?*" (Beck, 2010: 254). This question provoked his initial interrogation and he sets out to discuss and answer it in these 8 theses. According to Beck (2010),

1. Up until now the discourse about climate politics is expert driven and elitist and excludes and neglects common people, societies, their opinions, voices and their interests. For the author, transformation of climate change politics will only be possible by including sociology, specifically environmental sociology.
2. There is an important background assumption combined with general ignorance about environmental questions. Paradoxically this has been incorporated into environmental sociology. The pertinent criticism the author makes is that the category of *the environment* only includes what isn't human or social and the concept is sociologically empty. However, if the concept includes human action and society, it is scientifically mistaken and politically suicidal.

[4] Ulrich Beck – May, 1944 – January, 2015

3. Inequality and climate change are two sides of the same coin. It is no longer possible to conceptualize inequality and power without taking into consideration the consequences of climate change, and it is no longer possible to conceptualize climate change without taking into consideration its impacts on social inequality ties and power.
4. Later in his theses, the author states that Climate Change exacerbates existing inequality between rich and poor, between the centre and the periphery – but simultaneously dissolves them. The greater the planetary threat becomes, the more evident it gets that climate change is hierarchical and democratic because it reaches rich and poor alike. However, we know this is only partially true. Climate Change is pure ambivalence. It also launches a 'cosmopolitan imperative' – cooperate or fail! This can translate into the reinvention of green politics.
5. In the fifth thesis, the author highlights that regulation begins earlier and more profoundly with the question of how we overcome organized irresponsibility and have accountability, compensation, and proof. What for Marx were *relations of production* in capitalistic society, are for the risk society *relations of definition*. Both have to do with relations of domination. Among relations of definition are the rules, institutions, and capacities that specify how the rich should be identified in particular contexts (for example, within the nation-states, but also in relations between the nation-states). These relations form the epistemological and cultural matrix of legal power in which risk policy is organized.
6. The political explosiveness of global risks is due to how they are represented in mass media. When presented in the media, global risks can become cosmopolitan events. The presentation and customization of manufactured risk makes the invisible visible and creates simulations, tone, involvement, and shared suffering – doing so creates relevance for a global public. In this way, cosmopolitan events, public and private experiences, communications, kneejerk reactions and strokes of fate are highly selective, variable and symbolic both from the local and global point of view.
7. Finally, in his last two theses Beck emphasizes that ecological risks themselves will trigger the death of environmental politics. What is his position then about modernity and economic growth? Does modernity represent a sin against nature, or does it represent the courage to invent and open the way for an alternative modernity?

An alternative modernity would have to include a new vision of prosperity that isn't defined by the economic growth accomplished by those who worship at the altar of the market. *Wealth, instead of being defined in gross economic terms, would be defined in terms of general well-being.*

8. The Cosmopolitan is not only an urgent political and moral question, it is also a multiplier of power. Those who think only in national terms are the losers and only those who learn to see the world through cosmopolitan eyes can avoid decline and discover, experiment, and acquire new options and power opportunities that make a difference. The feeling of emancipation and power that emerges from overcoming national barriers is what can – potentially – awaken enthusiasm for the greening of modernity.

Hulme (2010) defended in his essay the fecundity of Beck's idea of cosmopolitan for understanding changes in the political, sociological, and psychological aspects of climate change. This argument was illustrated through brief examinations of how climate change is contributing to the dissolution of 3 modern dualisms: nature – culture (ontology), present – future (epistemology), and global – local (geography). Not only does the cosmopolitan perspective help us understand the ways in which science and society are mutually constructing the phenomena of climate change, it also offers us an intriguing way of questioning: What can climate change do for us? Instead of, what can we do for climate change? The author is as provocative as Beck when he says that it's sociologists who are needed in order to be able to answer this question.

In his article, Swyngedouw (2010) questions the relation between two themes that are seemingly disconnected: Consensual presentation and integration of the global world and the problem of climate change, on one hand – and the theoretical political/philosophical debate centred on the emergence and consolidation of a post-political and post-democratic condition, on the other. The author argues that climate change appears to be more politicized that ever and is being emphasized in the political agenda. On the other hand, various increasingly influential political philosophers insist that post or depoliticalization of the public sphere parallel and intertwined with neo liberal processes have been the fundamental indicators of political process in the last decades. The author highlights four stages:

- First, briefly we delineate the basic outlines of the argument and its mechanisms.
- Second, we explore the ways in which the present climate enigma is predominantly taught through mobilization of apocalyptic imaginings.
- Third, we argue that this specific representation of climate change and its associated politics, is sustained by populist gestures.
- Lastly, we discuss how this specific choreography of climate change is one of the arenas in which a post political picture and post democratic political configuration have been negotiated.

In Parks and Roberts article (2010), the authors sought to answer why North – South climate negotiations have gone on for decades without producing any substantial results. We argue that the growing lack of convergence on climate change is almost inevitably due to the world's inequality. This inequality has created and perpetuated highly divergent ways of thinking, visions of the world and causal beliefs. It has also promoted notions of particularized justice. The structuralist insight that uncontrollable inequality undermines cooperation, suggests that climate negotiations should be broadened to include a series of apparently unrelated development issues like trade, investment contracts, debt, and intellectual property rights. On reviewing the work of some *normative entrepreneurs*, we concluded we need to bring questions of justice to climate negotiations. We need to explore how these ideas might influence discussions about *burden sharing* in a post Kyoto world where development is limited by climate change.

In his article, Josanoff (2010) argues that because climate change produces disagreements on the ways we understand humanity's place in nature, it offers challenges and opportunities for the interpretive social sciences. Scientific evaluations like the IPPC have helped establish climate change as a global phenomenon. However, this occurred through a process that separated knowledge from meaning. Facts about climate emerge from impersonal observation, whilst meaning emerges from experience. Climate science therefore contradicts common sense and undermines existing social institutions and ethical commitments on four levels: community, political, spacial and temporal. The article explores the tensions that arise when the impersonal, apolitical and universal imaginary of climate change projected by science comes into conflict with the subjective, positioned, and normative imaginations of human beings involved with nature. It also points out the current environmental debates in which a reintegration of scientific climate representatives with social answers, is occurring. It suggests how social interpretive behaviour of science can promote a more complex understanding of the dilemma between climate and humanity. An important aim of this analysis is to offer a structure in which to think about the human and the social within an atmosphere that seems to have made the previous categories of solidarity and experience obsolete.

Finally, Shove (2010), makes and interesting contribution about the ways in which environmental social theories and climate questions have changed over the last year, what topics have been concentrated on and where dialogue has broadened. The author comments that many of the articles included in this special edition exemplify the tendency to frame climate change problems in terms of existing concerns, such as the character of capitalism; the relation between nature and culture; or the social problem definition process. Other forms of conceptual development are more obviously directed to the challenge of understanding, and possibly promoting, social transformation in response to climate change. Meanwhile politics continue a characteristically narrow account of the social world. In the article, the author tries to highlight the differences in the ways these agendas unfold, and comments on what this means to the types of questions social theories have often asked, and have yet to ask, about climate change. In conclusion, social theory – broadly defined – has a lot to offer, but reaching its potential requires concerted effort and active involvement with new and unknown audiences.

It is also worth analysing the relationship established between climate change and the Anthropocene as described by Chakrabarty (2017). The author states that the global climate change discussion has been moulded by intellectual categories developed to tackle capitalism and globalization. Nonetheless, climate change is only one manifestation of the diverse and accelerated impacts of humanity on Earth's systems. The Anthropocene raises questions about the distribution of justice between the rich and poor, developed and under developed nations, between the living and unborn – and even between human beings and nonhuman beings. It challenges the categories in which our traditions of political thinking are placed. For the author, the awareness of the Anthropocene encourages us to think about human beings in a different context – as parts of a global capitalist system and members of a now dominant species. At the moment, however, this debate is built on the concepts and experiences of the developed world.

Cielemecka and Daigle (2019) claim we are being confronted by an environmental crisis induced by mankind on an unprecedented scale. This calls for a new theoretical approach that abandons human exceptionalism and concentrates on developmental projects and ethical questions that are suitable for our times. The authors offer an anti-anthropocentric project as an ethos for living in the Anthropocene through revisiting the idea of sustainability in order to problematize humanity's linear vision of the future, and the uniform 'we' of humanity on which it is based. Their analysis, anchored to concepts of post humanism and materialist feminism, seeks to establish a dialogue that offers a concept of post human sustainability. One that decentralizes the human and repositions it within its ecosystem, while remaining aware of differences, and promoting the prosperity of all walks of life.

Skrimshire (2019) argues that the concept of the Anthropocene permits human history to be imagined within the time and structure of planetary processes. Some environmentalists increasingly favour massively stretching time horizons of moral concern. In the first section, the author presents two recent examples of the idea of long-term vision: (a) narratives about the future in popular climate science and in futurism; (b) the ideas behind the Long Now Foundation. In the second section, he takes a critical look at these perspectives through the classical analysis of secular time, and suggests that in order to avoid new materialist perspectives, this post secular criticism should be considered alongside recent approaches to the Anthropocene and to geology.

This revision of thought recognizes the radical, analytical thinking dedicated to the criticism of capitalism and its impacts on the environment. The author, Mary-Jayne Rust, presents two important studies – one in 2004 and the other in 2008. The author is an Ecopsychologist, Jungian Analyst, and Art Therapist. She began her work in 2004 by asking herself how psychiatry needed to change in order to help create a sustainable future. With the aid of descriptions from indigenous cosmology, she examines the meaning of sustainability, and explores the concept of the self in relation to nature and culture. The article asks how we identify ourselves with the greater whole and why we have disconnected from it. It suggests that psychotherapy is a powerful tool for reconnecting with the world. There would be benefit to expanding beyond the human sphere and embracing our interests and relationships with the non-human world.

This would involve relating to nature as a subject, and embracing our anthropocentrism. The article questions how this can take place in both our everyday life and in our inner world.

In an article in 2008, Rust presents a revised version of a talk given on November 17, 2007, at the Annual Conference of the Psychiatric Union in London, UK. The aim of the talk/article was to explore the personal psychological attitudes underlying climate change and the ecological crisis. The central question was if psychological insight could contribute to the collective change needed to move towards sustainable life.

- The first part explores two great myths that sustain western culture: the myth of the fall and the myth of progress. Our understanding of these stories has kept us imprisoned in destructive ways of life. Western culture in particular has developed a long time internal and external fear of wilderness. This is in contrast to an indigenous world view where humans respect the need to maintain a balance between human beings and the rest of nature. How can we find a way of working with nature in modern times?
- Part two explores our personal fantasies and answers for a sustainable life. Consumption became the opium of the people in order to subjugate our inner wild nature. Such an addictive relationship keeps us from thinking and acting. Recuperation involves replenishing our bodies by developing what Naess calls an ecological identity.
- Part three explores how these questions can enter our work as therapists and how we can respond.

Randall (2005), who we have already mentioned and who often appears in our work, is an analyst who has developed work related to the environmental crisis and climate change. In this article she analyses that psychotherapy is a cultural practice subject to the broader ideologies of the time. She asserts that psychotherapists need to understand how this manifest in the individual psyche. Psychoanalytical understanding of unconscious processes can contribute to comprehending the difficulties environmental activists face and why people resist the environmental message. Contrary to appearances, anxiety over climate change and environmental damage is acute but defended against through primitive psychological processes on a collective level. This article explores various aspects of this process including the difficulties that the processes of division, projection, and infantilization produce for people in the environmental movement. There is a tendency to guilt resulting in a repression of desire for objects of consumption, that is inevitable and exacerbated by a projection of split guilt on the rest of the population. This projection lodged in the environmental movement and was later attacked and resulted in the commonly known caricatures of environmentalists.

A curious and provocative article by Fay (2016), briefly analyses the past history of psychotherapy and its future perspectives. It suggests that, although psychotherapy may continue to aspire to save the planet 'one person at a time', it also needs to broaden its efforts and work with a greater number of people on a broader playing field. This arena would include all urgent social issues together with the political, cultural, and ecological questions of the new century. This requires a new politics of indigeneity that recognizes locality and localism, and not only profoundly respects the unique contribution and leadership of the indigenous populations of First Nations, but is also bold enough to claim that all of us are indigenous to the universe. Because of relationship, we share the responsibility of learning to be better caretakers and caregivers of ourselves and each other on the *Garden Planet*.

In House's (2019) manifest article, the author is explicit in the affirmation that life is dying in the interdependent, complex, global ecosystem in which we live. Through accelerating species extinction, the climate crisis is moving faster than was previously anticipated. In the opinion of the author, a global catastrophe is inevitable. Especially if political leaders around the world continue speaking about the climate crisis, at the same time corporative global capitalism continues to propel the international economy. Complacency and inaction in Great Britain, The United States, The European Union, Australia, Brazil and throughout Africa and Asia show diverse manifestations of political paralysis and the abdication of humanity's critical responsibility for managing the planet.

In the vision of the author, national and international political organizations must immediately prioritize the question of climate emergency and urgently set forth wide reaching policies to resolve it. Conventionally privileged nations should voluntarily finance impoverished nations, and compensate them for resources by renouncing their unsustainable economic growth and imperialistic ransacking of the planet.

In the authors critical view, the effects of extreme climate shifts have already reached food production. With this comes the risk of extreme hunger and need for emergency investment in agro-ecological production of food crops resistant to extreme climate conditions. We also call for an urgent summit to examine options for saving the Arctic ice cap. This would reduce climate disturbances to our crops. The author's text cites the importance of concerned global citizens standing up and organizing within their spheres against this present complacency by defending the rights of indigenous peoples, decolonization, and restorative justice. Collectively, we must do what is necessary, non-violently, to persuade politicians and corporate leaders to renounce their ignorance and complacency. *Business as usual* is no longer an option. Global Citizens will no longer put up with this failure of our planetary responsibility. All of us, especially those of us in the materially privileged world, need to commit to a lighter foot print. We must consume far less and not only defend human rights but carry out our responsibilities as caretakers of the planet.

In 2016, Rekret presented an article that critically evaluated the conception of the underlying ethics of the growing constellation of social theories known as *new materialists*. The author argues that these theories are of little or no importance to understanding contemporary transformations of the relatedness between mind and body, or between human and nonhuman natures. Using the work of Jane Bennett, Rosi Braidotti, and Karen Barad as an example; this article claims that continuity between ethics and ontology is central to recent theories of 'materiality'. These theories affirm the primacy of matter, calling for a spiritual or ascetic self-transformation in order to 'tune in' or 'register' materiality. Inversely, portraying criticism as hubris, smug, or resentful and blind of its anthropocentrism. It is argued that framing the fundamentals of ontological speculation in these ethical terms, gives permission to omit analysis of the social forces that mediate the access of thought to the world. This allows the theoretical to bypass any questions about thought conditions. In particular, the paper points to processes of so-called primitive accumulation underway in forming the relatedness between mind and body, human and non-human.

Seixas and Nunes (2017), discuss the period of unstable experimentation together with the challenges of globalization and associated risks; the disenchantment with 'lasting injustice'; and in studying subjectivity consider environmental changes and mental health. We begin by identifying how the principal studies about climate change and mental health are incapable of explaining the emerging and co-evolutionary paths of agency. As a way of freeing these studies from their objective dimensions and linear causality, we argue in favour of repositioning subjectivity within a framework that recognizes conflicts beyond the super deterministic determinations of the centres of power – state, market or religion. We base our discussion on an example of scientific research that was carried out in a region of intense environmental, social and cultural change in the state of São Paulo, Brazil. The goal is to open discussion regarding research on mental health and we have contributed the notion of the pluriversal subject to these discussions.

In the following we give special emphasis to the authors who have dedicated themselves more intensely to the analysis of climate change and subjectivity, as a well completed example of cultural narratives about environmental problems. We also want to emphasize their self-representation as psychoanalysts.

To Mnguni (2010), sustainability is thought of as a quest for meaningful coexistence within, and amongst, social systems (sociocultural sustainability); and between social and ecological systems (ecological sustainability). It is a search for a state of psychological maturity and integration whereby each person is, for the most part, able to demonstrate capacities for reality-based relatedness and for transcending the kinds of paranoid-schizoid splits that characterize the psychological functioning of an infant. Work for sustainability can be thought of as representing a search for individual and collective maturity – a search for both mature human-human relatedness and mature human to nature relatedness. Thus, developmental, and regressive dynamics in sustainability become inevitable and necessary. Inevitable because growth is never linear, and necessary because of the developmental opportunities that regressive moments, if meaningfully engaged, potentially represent.

The problem of sustainability is complex and multifaceted; it involves an intricate web of connectedness amongst psycho-social and ecological issues. This complexity places sustainability in the inter-organizational domain, making collaboration by multiple stakeholders' imperative. Partnering for sustainability is not easy, for it is replete with tension and contradiction. Tension in sustainability work derives from the inherently paradoxical nature of: (1) human–nature relations; (2) collective life and; (3) inter-organizational domain. This inherent tension sets the scene for heightened anxiety and for complex conscious and unconscious dynamics, which, if not adequately processed, can derail collaborative effort. They ultimately can render the project unsustainable (Mnguni, 2010).

While people generally focus on the positive aspects of sustainability – its creative and restorative intent – looking at it through a psychodynamic lens serves to bring to the surface some of its unconscious dimensions – its shadow side. Social defence theory helps bring into sharp attention the nature of the anxiety that attends the primary task of trying to restore socio-ecological landscapes; and the defensive routines used by sustainability workers to cope with this anxiety.

Splitting in sustainability is evident at several levels. First, while in the beginning all societies had closer ties with nature, with the advent of agriculture and then industrialization, some sections of society seem to have continued to enjoy and honour these close ties, while others seem to have gravitated towards the opposite end of this tension and opted for an almost complete estrangement from nature. It is as if living with the inherent contradiction between the needs of economic development and those of social and ecological sustainability, causes too much psychic pain. The bifurcation of human–nature relations then becomes both a form and a consequence of such splitting mechanisms. So, to Mnguni (2010), sustainability work is replete with tension and anxiety. However, unlike other helping professions in which the emotional toll of the work is generally recognized, this aspect of sustainability work is not sufficiently acknowledged.

It is, nowadays, common for governments to provide funds for projects that are then expected to solve complex societal issues, like sustainability, where governments themselves have failed – as if those working in complex problem domains have a magic wand. It seems reasonable to consider that those who fund sustainability initiatives may not have any real expectation of successful outcomes. Providing funding then becomes a gesture, so that sponsors can feel that they are – and are seen as – 'doing their bit'. It is also not unreasonable to suggest that those who work for sustainability, at some level, realize this. Hence, as articulated by a participant, feelings of futility can come to characterize people's experiences of sustainability work. Now and then, workers may be forced to confront questions about the real impact of whatever it is they might think they are doing.

As stated earlier, the primary task in sustainability – *to reverse or halt the escalating degradation of socioecological landscapes* – shares parallels with that of hospitals and other care facilities. Sustainability work can be thought of as representing attempts to cure societies of their social and environmental ills. To the extent that some aspects of this work are impossible, work for sustainability can be expected to exacerbate workers' anxiety. The complex and interconnected nature of sustainability issues is such that it is quite common for a solution in one area to trigger unintended consequences in other areas. For example, wind-power generation, a potential solution to global warming, has been known to have the unintended consequence of killing birds, including rare species.

Adams (2014), as mentioned before, analyses that we face a serious environmental crisis. One which poses a threat to planetary, social, and personal continuity – and it grows daily. In the authors perspective (Adams, 2014), recent work exploring the effective and social dynamics of defence mechanisms is one way of transcending a preoccupation with rational decision-making processes in sustainable behaviour. This gets us closer to a meaningful understanding of inaction in this context.

To Adams (2014), according to psychoanalytic theory, defence mechanisms are triggered by anxiety-inducing situations that require us to confront 'painful material' of one kind or another – perhaps information about ourselves or those close to us we find unpalatable. Défense mechanisms allow us to "*deny or pretend the problem is not there, or that it is the responsibility of someone else*". Attention is paid to climate change denialism, like campaigns of misinformation about climate change funded by commercial and ideological interests, and it is also paid to everyday and unconscious forms of denial. Adams (2014) sees denial as a socially organized and individually experienced effective response. It is the flesh on the bones of canonical cultural narratives, geared towards 'business-as-usual'.

Adams (2014) mentions that, concerning sustainability, there are two parallel narratives relating to the climate change discourse: One frames the problem, the other the solutions. This is an exemplary case in point as mentioned by Randall (2009), that ranges from the loss of biodiversity, species, crops, water, and fuel, catastrophic loss is central to the narration of the problem. Solution narratives, on the other hand, do not tend to acknowledge loss at all but they present simple palliatives as solutions in the present. Examples include 'small steps'– suggestions that individual lifestyle changes on a mass scale are sufficient to avert crisis; calls for ethical consumption – transformation through the consumption of green products; and faith in technology – life can go on as normal once low-carbon alternative energy sources, forms of mobility, etc. are discovered.

The potential solutions offered by diverse authors, according to Adams (2014), vary. However, it is generally agreed there is a need to confront the reality of loss associated with ecological degradations, so that the energy currently invested in denial can be converted into something more positive. At the same time there is a tendency towards a reiteration of the general myth of progress in *business-as-usual* framing of how to respond to environmental crisis. Business-as-usual narratives pin their hopes on technological fixes that will allow us to continue our lives more-or less as we do now – lives informed by a narrative of progress. On the other hand, 'apocalyptic anti-futures' speculate destruction and annihilation in the near or distant future.

Another aspect to consider, as mentioned by Adams (2014), is that in many cases, communication regarding our role in ecological degradation generates uncomfortable responses, while it challenges the narratives we live by and our identification with them. This discomfort triggers defence mechanisms, such as splitting and projection. They take narrative forms that creatively reflect socially organized and cultural ways of framing our response (including the assiduous promotion of denialism). This, in turn maintains, reinforces, or at least blunts any challenges to those narratives we live by, and the associated feelings of responsibility.

This discussion, according to Adams (2014) calls for a deeper and broader understanding of the psychology of sustainable consumption and pro-environmental behaviour, including associated challenges. In developing that analysis, defence mechanisms are seen to be an important way of approaching the emotional complexity and inconsistency of apparent inaction; this is in keeping with the context of increasingly sophisticated communication about ecological degradation and the contributing role of human behaviour. In widening our analysis, the discussion has pointed to accounts of the social organization of denial. Here the narrative frames denial and other defence mechanisms, as social psychological processes, operating at the level of individual and intersubjective experience. They are established and broadcast socially and interpersonally.

In his paper Adams (2014) has claimed that narrative frames are an important dimension of the social organization of denial. Narratives are understood to be a universal and fundamental vehicle through which human life is made meaningful; yet narrative frames rely on indexed, historically contingent social norms; and established power relationships that force and facilitate identities into particular shapes. In contemporary consumer societies, canonical cultural narratives, (specifically those articulating environmental problems and sustainability), play a key role in organizing denial. They provide us with culturally validated opportunities to minimize loss or project it into the future; blame others or project responsibility onto them; and encourage us to consider consumerist lifestyles, as a potential solution to ecological degradation.

In her article on Renee Lertzman's, *Environmental Melancholy*, Adams (2017) highlights that human engagement with ongoing ecological crises, are predominantly defined by loss, melancholy, and ambivalence. Lertzman's subsequent argument, according to Adams (2017), is that an apparent unconcern or indifference about environmental issues is a mask – a way of shielding us from deeper anxiety, loss, and concern. Indifference can also shield us from a desire for reparation: to express care, compassion, and make amends.

Adams (2017) mentions that ambivalence, in Lertzman's perspective, is apparent in the way people talk about the development of local industries. It vacillates between positive and negative. Ambivalence is also present in what is not said – the tendency to skirt around it as one might protect a problematic but beloved relative. 'Beloved' is an apt word here because the ambivalence Lertzman witnesses, as mentioned by Adams (2017), is also a manifestation of her respondents' effectual investment in local industries as objects of progress and development, and as providers of employment, identities, and associated securities.

According to Adams (2017) analysis and comments, industrialization, and our experiences of it, are integral to the object relations Lertzman describes – a socially, culturally, and historically contingent set of dynamics central to the psychodynamics going on 'beneath the surface'. In adopting an object relational framework, she interprets her participants' talk as manifesting a network of relational dynamics, that reaches 'outwards' towards the systems and structures of 'industry' and its representatives; as well as 'inwards' towards their own effectively invested impressions and identifications.

Ambivalence, in other words, reflects the highly complicated nature of our deep investments in practices that are both life affirming and life-degrading. Thus Lertzman, according to Adams (2017), defines environmental melancholy as a condition in which, even those who deeply care about the well-being of ecosystems and future generations, are paralyzed in translating such concern into action. So, developing opportunities for people to make meaningful reparative contributions is vital for mobilization, and has the potential to radically reframe our strategies, tactics, and orientations to environmental engagement.

Adams (2017) mentions that Rosemary Randall has put strong emphasis on the need to create support structures that facilitate the process of mourning and provide containment for the anxieties that will inevitably be revealed. We need strategies that deal with the difficult issues of status and identity, and a culture of stories and role-models that offers meaningful examples to identify with. In this perspective, Adams (2017) mentions that one approach presented by Lertzman, could be to focus on ways that may enable people to experience themselves as true agents. Agents with opportunities to contribute, create, be heard, respected, and valued. Also, there is a need to address communication strategies that deny or distract us from working through loss, anxiety, and ambivalence; and develop alternatives that rhetorically acknowledge emotional responses relating to fear and loss or encourage creative participation in environmental engagement.

Another aspect of Adams' (2017) analysis is the active resistance by regime actors to promote fundamental change. This source of inaction contributes to the ecological crisis and the ambivalence of our citizens. These actors are key business and government players, oriented towards maintaining the status quo. They forge 'core alliances' through mutual interests, positions of authority, access to media, common experiences, and worldviews consolidated by regular proximity and shared spaces. This notion of *regime resistance* is supported by analysis of strategic denialism, which refers to intentionally coordinated campaigns of misinformation, funded by commercial and ideological interests. These campaigns disseminate information that is explicitly designed to unsettle the anthropogenic facts of climate change; question its anthropogenic origins; and/or question our ability to meaningfully intervene.

In Adams' (2017) perspective, *environmental melancholy* shows us that we are as profoundly invested in a degraded natural world, as we are in the social, cultural, political, and material structures and systems, from which this situation has emerged. In addition, Adams (2017) says that Lertzman makes a valuable and significant contribution to understanding the contradictory and complex nature of building a movement against ourselves.

To Lertzman (2012) a rarely acknowledged, fundamental ambivalence and contradiction, is arguably at the heart of ecological damage. Without explicit acknowledgment, this conflict tends to become internalized, and to be based on environmental discourses. Lertzman (2012) also says that we face this through parallel narratives of catastrophic loss and idealized solutions. She says that environmental studies need to weave into their discourse the recognition of human desires, fears, anxieties, and hopes – not only what our attitudes, beliefs, values, and opinions may be, or which smart device we use to turn off the lights. Environmental dilemmas are ripe for deeper, more nuanced understandings, and approaches. What is needed are solutions and work that can incorporate conflict, anxieties, losses, melancholy, creativity, and the desire to make reparations in the world.

In this perspective it is relevant to analyse Ben-Asher and Goren's (2006) paper that examines a unique way of contending with a life-threatening ecological hazard. The article examines the psychological use of projective identification to deal with the dangers inherent in a severe ecological threat. According to the authors, although the ecological threat of environmental pollution is not perceived as a local problem and there is ever increasing awareness of the issue, attitudes toward ecological threats are generally repressed and distanced from the lives of individuals. Instead, such problems are assigned to the field of operation and responsibility of 'green' organizations or unique groups of 'environment-freaks'.

The paper presents an incident in which soldiers who served in an elite commando unit in the Israeli Army (Israel Défense Forces – IDF) had, for years, been exposed to a threat to their lives due to training in the waters of a river polluted with heavy metals. A concentration of heavy industrial plants, including refineries and other petrochemical factories, was built in the vicinity of Haifa, near the point where the Kishon River flows into Haifa Bay and its large commercial port. The waste from factories, contains carcinogenic and mutagenic materials originating from heavy metals that do not break down. High concentrations of these materials cause various diseases, such as cancer, in humans who are exposed to them. Acknowledgement of the danger and the subsequent demand to treat the soldiers who were already sick, as well as those who were not yet sick, occurred by means of dramatic changes in the relationship between the soldier-sons and their parents. The actions of the parents assumed unique psychological significance because their soldier-sons projected their anxieties and could therefore remain loyal to the military unit while apparently objecting to the organization and protest of their parents. The principal concept employed to examine the relationship between the parents and the soldier-sons was projective identification, which serves as a psychological process that is simultaneously a type of defence mechanism, a means of communication, a primitive form of object relationship, and a pathway for psychological change. After a transformation has occurred, the sons can take their projections back from their parents, the object with which the concern was deposited for safekeeping. The authors concluded that it was too soon to know whether the social lesson has been learned regarding the value of human life. It is unclear whether the responsibility accepted by the government to care for the sick soldiers, will simply serve to placate the population of parents; who will thus discontinue their struggle to change the situation and prevent future disasters; or whether it will constitute a genuine change in perception, in which ecological risks are regarded as one of the potential dangers to the lives and welfare of civilians and soldiers alike.

According to Brown (2016), humanity has known for a long time the consequences of pollution and environmental degradation – as though following some tragic historical necessity but being unable to respond. At the same time, problems like overpopulation, climate change, depletion of the Earth's natural resources, and landscape changes, all engender feelings of sadness, anxiety, and concern. A pervasive sense emerges that something needs to be done; yet this impulse is perhaps at odds with an attitude of open curiosity when there is no action being taken.

To Brown (2016) the way many have attempted to deny the reality of climate change, pushes those with a belief in this notion towards a polarizing reaction. This instigates a pervasive tone of authoritarianism, which some have even come to argue may be politically necessary. Therefore, emphasis is placed on convincing others of a truth which is regarded to be incontestable. While reducing carbon emissions seems critically important, emphasizing solution-based responses of this kind may in a sense run counter to a more substantial change.

Some authors, according to Brown (2016) emphasize a tendency to approach the environmental crisis in terms of the general population's apparent inability to address it. Hence, psychoanalytic theorists interpret the crisis in terms of a failure to act, and draw attention to, such themes as our collective anxiety, apathy, denial, or destructiveness.

Analysing the reflection of Jungian analyst V. Walter Odajnyk on the threat of an atomic war, Brown (2016) mentions that Odajnyk articulates that people prefer to deal with this threat as they do with their personal death – they tend to ignore it or leave it up to fate. Such attitudes may be valid and even beneficial with respect to individual life, but they are inappropriate and dangerous to the life of the species.

Most of us, according to Seppänen (2011) are aware of climate change but, at the same time, are unwilling to change our actions or even think about the whole issue which seems too frustrating, complicated, and distressing to bear. Hence, denial is paradoxical – it is both knowing and not-knowing. The cultural repercussions of denial and repression extend their influence from the ordinary flow of everyday life to the most terrible historical events. Sometimes these two overlaps, politically load the unconscious mind. Therefore, overcoming denial and repression opens a possibility – but only a possibility – of facing up to and acknowledging socially and politically loaded environmental problems.

Hoggett (2011), says that climate change is a classic example of 'post-normal science', where facts are uncertain, values in dispute, stakes high, and decisions urgent. Lacking predictive powers, we simply do not know how objectively bad the situation that we are confronting is.

The growing recognition of the reality of anthropogenic climate change, challenges us with the same collective psychic predicament: How can we think in a realistic way about something whose implications are unthinkable? Given the nature of climate science, we just do not know how bad things are likely to be, and, given the uncertainty, all we can do is try our hardest to make sure we avoid the worst.

To Hoggett (2011) the situation is alarming. And, whilst complacency is no longer an option, neither is a politics of climate catastrophe a viable alternative. Catastrophism is a politics of despair that draws on the same survivalist mentality as previous apocalyptic movements did. Survivalism is a demoralized state of mind in which questions of value have been progressively destroyed. Its radicalism can quickly take on the authoritarian cloak of lifeboat ethics, but without passion, a politics of climate change is disarmed. We need to be thoughtful, generous, concerned, sceptical, and able to manage uncertainty; but we also need to yoke this to a fighting spirit, to anger, hope, and yes, at times, to unreasoned passion.

Samuels (2015), analysing Naomi Klein's, *This Changes Everything: Capitalism vs. The Climate*, says that it presents a careful analysis of why the fight over climate change will have to take on our current system of capitalism; and why this fight has so far failed. The author (Samuels, 2015), also highlights the idea that at the very moment that scientists were confirming that only human intervention can reverse our march towards an uninhabitable environment, the neoliberal political agenda was gaining full force. This new conservative ideology argued against government, taxes, public programs, and science-based public policies; and helped undermine any belief in the ability of our governments to take on the huge, long-term problem of climate change. In fact, the neoliberal free-market ideology has pushed governments to double down on their production of fossil fuels, at the precise moment when it is necessary to replace fossil fuels with alternative energy sources.

On a psychological basis, Klein, according to Samuels (2015), makes a convincing argument that humans have a hard time with long-term thinking. So, it is difficult for us to take on a problem like climate change, which requires balancing short-term interests against problems that will arise in the distant future. It is hard to argue with this common human tendency, but psychoanalysis requires us to go further and ask why people knowingly participate in their own present and future self-destruction. Our first psychoanalytic intervention into climate change then tells us that appeals to people's self-interest may not be enough to persuade people to stop participating in a system that is projected to destroy our way of life. If people enjoy painful experiences, and they would rather repeat instead of changing destructive behaviours, our normal ways of doing politics will not work. The death drive and unconscious masochism, undermine our desire to see politics on a level of rational self-interest. If people enjoy their self-destructive behaviours, they may not mind if they are participating in the devastation of our environment.

Returning to politics, on a fundamental level it does not matter if conservatives are to blame for climate change or our failure to fight it; what matters is how we *all* participate in this system of self-destruction and whether we can give up the fantasies that prevent us from fixing the problem. However, since liberals want to see conservatives as the perpetrators of all our problems, liberals do not have to deal with their own role, and so liberals remain innocent victims of climate change. It turns out that the central problem of psychoanalysis is also the central problem of global politics and the future of our world. How do you get people to give up the self-destructive fantasies that guide their perceptions and beliefs? Part of this problem concerns what allows people to move away from actions and thoughts that stabilize identity and bring unconscious enjoyment through suffering.

2.6 Some Important Considerations About Objective Advances in the Relationship of Subjectivity, Psychoanalysis, and Climate Change: The Construction of Networks

In this section we highlight important aspects that help bring deeper understanding of climate change to society. It also highlights the subjective ways consciousness is being raised and how climate impacts and problems are being faced through building networks and alliances. We believe that networks can be important instruments of change.

The networks proceeded from an important narrative in psychoanalytical literature contextualized and consolidated by a group of research scientists. This group has an existential concern about the environment and the degradation of natural resources specifically related to the impacts of climate change on nature, the economy, and human subjectivity. Through their theoretical approaches, they have analysed the role of denial of these problems in society. Many of these researchers spin-off from specific points present in several of their previous studies, and place special emphasis on the following:

1. Climate Change is recognized as the great catastrophe of the future of humanity and the planet. It includes the present and future devastation, and also the complex dimension and extensive range of unfolding catastrophe, that initiates extreme events that damage human health, biodiversity, food security, and creates masses of climate refugees.
2. The recognition and fear of climate change for present and future generations imposes changes to behaviour and social values (IPCC, 2014; Tschakert et al., 2017; Kabir et al., 2016; UN Environment, 2019).

It isn't our intention to cover all the networks that are presently being built or have already been built. For this work we have chosen as examples two networks, *The Climate Psychology Alliance* and *Cambridgecarbon footprint.org* (CCF), whose members' priority centres on psychoanalysis. These networks, located in the United Kingdom, have made a powerful contribution to the reflection on climate change and its relation to subjectivity and psychoanalysis, from a theoretical standpoint; and they also contribute to the question of the social support needed for climate change impacts.[5]

---

[5] Other examples of networks that can be consulted are the International Community for Ecopsychology, at https://www.ecopsychology.org/ and #Manchester Climate at https://manchesterclimatemonthly.net/about/our-missions, which despite being active and interesting, are beyond the scope of our choice in this work.

The Climate Psychology Alliance (CPA)[6]
The CPA is a non-profit organization focused on the connections between deep psychology and climate change. It is mostly made up of psychotherapists but is also open to other professionals that find this perspective useful. It is based in the United Kingdom and there are many interconnections with branches now extending to the United States. It emerged in December 2009 after many conferences and meetings that took place at the Centre for Psycho Social Studies (CPSS) at the University of the West of England. These events helped mobilize and bring together people with combined backgrounds in psychotherapy and academic work. Its members are greatly concerned with the evidence of climate and ecological destabilization resulting from human activities.

Since its founding, the CPA has organized a series of events, workshops, and publications directly related to climate change.[7] These events and workshops have brought psychoanalytical and psychosocial perspectives together to collaborate on the challenges of climate change, ecological crisis, and the role of human engagement. It has uncovered challenges and questions for the group that are directly related to the importance and roles of diverse areas of psychology, and their capacity to contribute to research science, the government, and the public sphere. Climate change is the greatest challenge to the survival of the planet and to all species and is the greatest problem that has ever been created and faced by humanity. We as a group are motivated to transcend our cultural differences for the greater good – bringing together energy, ideas, and potential influence. Meeting these challenges is at the heart of building the CPA network. The CPA also seeks to contribute to the task of mobilizing a relevant collective response involving the work of mitigation and adaptation.

The CPA created the term *Climate Psychology*. This term is being more and more widely adopted. The very definitions of the CPA reflect its roots in depth psychology. This was established (2018–2019) in its Manual of Climate Psychology.[8] Reading its publications, which include the manual, workshops, podcasts, papers, blogs, and a social media presence on Facebook and Twitter, etc., shows that the story of the CPA is remarkable for its involvement in various themes such as negation, ecocide, radical

---

[6] For more information see: https://www.climatepsychologyalliance.org/

[7] In 2009, the first event was "Facing Climate Change" conceived and chaired by Adrian Tait, with speakers George Marshall, founder of Climate Outreach (formerly COIN), Professor Paul Hoggett, Director of CPSS and ecopsychologist Mary-Jayne Rust.

[8] Handbook of Climate Psychology, Available: https://www.climatepsychologyalliance.org/handbook

hope, leadership, narrative, threats, uncertainties, and the emergence of the Anthropocene. Despite its pluralist ethos, a psychosocial perspective has been a key ingredient in the focus of the CPA. This emphasis on the interaction of social and psychological processes has allowed for the construction of a powerful set of tools to help understand the multiple environmental problems humanity now faces. Some examples of problems that call for this dual perspective are: the various forms of denial; the notions of consumerism and wellbeing; right to think; and gaps in political engagement with climate and the ecological emergency.

The publication of the IPCC report in 2018 triggered an interesting public reaction. The increased awareness that came about after the publication, plus the of media's coverage on the loss of biodiversity and symptoms, (such as the increase of extreme climate events and commentaries that highlighted the unsustainability of human demands on our planet), generated a good deal of societal anxiety.

In addition to the CPA's community practice of disclosing the psychological impacts of climate change on society, the association established links to initiatives that build awareness in civil society about climate change and the urgency surrounding it. Some examples of these movements are Extinction Rebellion[9] and Deep Adaptation.[10] The CPA seeks to bring human and natural sciences closer together to gain understanding of life that is ecologically viable, and to seek ways to support public ability to face difficult truths. The network is built on the perspective that *climate change isn't a scientific problem in wait of a technical solution – it is an urgent, terrifying, systemic problem involving environment, culture and politics.* Climate change generates fear, denial, and despair in individuals; and evasion, indifference, and duplicity in systems of governance. It imposes awareness about uncomfortable problems with justice systems, abuse of nature, and equality amongst peoples. The most provocative thing for the group's members to look at, is the fact that climate change challenges each

---

[9] See more in: https://extinctionrebellion.uk/
[10] *Deep Adaptation* is a concept and social movement created by Jem Bendell, professor at the University of Cumbria, 2018, based on the view that humanity needs to prepare for the possibility of societal collapse, as environmental change increasingly disrupts social, economic, and political systems. Unlike climate change adaptation, which aims to adapt societies gradually to the effects of climate change, Deep Adaptation is premised on acceptance of impending abrupt transformations of the environment. See more: https://deepadaptation.ning.com/

and every one of us in modern society from both a personal and political perspective.

The CPA's way of working with all of these issues is based on a wide range of perspectives that include philosophy, the arts and humanities, and systemic thinking. The principal focus, however, is on psychosocial studies in the field of psychotherapy. Approaches that help to understand unconscious, emotional processes that control thoughts, beliefs, and behaviours that manifest in societal systems of defence which mutually reinforce each other. One valuable observation that was presented on their site is that, "*Anxiety, guilt, and shame make it very difficult for people to face the reality of climate change. At the same time, those emotions lead to a denial that is validated and reinforced by the norms and structures of daily life.*"

Five key principals define the CPA's approach:

1. Recognition of human participation in the creation of environmental problems – because in order to reach a solution we first need to recognize that we are part of the problem. Some of the richest 20% of the world's population have lifestyles based on heavy fossil fuel consumption. To confront this, the CPA aims to challenge cultural norms of privilege and resentment. This is a radical strategy that eliminates greenhouse gases from the global economy. The CPA aims to challenge the consumerist paradigm which includes the notions that humans are superior and separate from nature; the idea that all of the problems we create can be resolved by technology; the dangerous illusion that the economy is of a higher order than ecology and that environmental costs can be 'externalized'.
2. The second principal is concerned with what is called 'existential shame'. This would be the slow awakening of awareness of our destructive exploitation of natural resources. This coincides with a growing awareness of the Earth as a living being, (Gaia). It is interesting that an analytic vision sees this as an attack on 'the mother that nourishes our species', but whom we have come to resent as a function of our modern technological arrogance and feelings of dependence. Climate psychology tries to approach these existential feelings of shame as a kind of sense of self-betrayal. The repair of this rupture, this alienation from internal and external nature, is an enormous challenge. Our species is facing a rite of passage marked by the end of a civilization moved by carbon.

3. The third principal suggests that it is important to secure the tension between hope and despair and that, instead of separating these connections, we need to maintain them. Keep them between complacency and alarmism; and between action and reflection. When these tensions are successful a vision that attracts creative and practical engagement can emerge.
4. The fourth principal is to offer understanding and support. Climate psychology recognizes that anxiety and feelings of abandonment are generated when attempting to face the terrible possibility of self-destruction and the possible extinction of species. It seeks to understand denial, as well as to give emotional support in dealing with damaging psychological impacts.
5. Lastly, the fifth principal is to restore what was repressed by recognizing that technological short-term success in society offers a false promise of endless growth. This falsehood breeds a complacency that seemingly justifies the domination and pillage of Earth's finite resources. Climate psychology therefore seeks to restore what has been repressed. It welcomes feelings and brings values and justice back to the human image. We recognize this successful project as neither an absolute belief, nor the only truth.

In the vision of the CPA, the recognition of these five key principals, through different approaches and in different degrees, allows the creation of a response to the climate anxiety that is gestating in humanity. This emotion is spreading and is beginning to permeate diverse therapeutic spaces. A curious statement on CPA's site says, "*Many of the 'canaries' of climate change, such as activists and scientists, at times must face uncontrollable feelings of despair, rage and sadness*", as is evidenced by the innumerable calls for support that the Climate Psychology Alliance is receiving.

The CPA network strongly believes, and we share in this belief, that the therapeutic community – united in this approach – has a vital role both as a support and for helping to deepen the understanding of how climate anxiety occurs both in our individual lives and in our culture. We need to prepare ourselves to understand and shine light on psychological resources that build the capacity to mourn these losses, (for instance, acceptance of the tragedy of the mass extinction of species), as well as build resilience, courage, radical hope, and new ways of imagining that allow us to find additional narratives and actions to assist in dealing with these changes.

Cambridgecarbonfootprint.org[11] (CCF)

Cambridgecarbonfootprint.org (CCF) is a registered charity that works to raise awareness of climate change issues and to support people in moving to low-carbon living. CCF seeks to stimulate the transition to low carbon communities which are sustainable, resilient, and rewarding. Its mission is to inspire people to engage with climate change and empower them, both individually and within their communities, to move towards low carbon living.

CCF offers a varied program of events and activities aimed at inspiring people to engage with climate change and empower them to build a low-carbon future. They seek to engage on many levels, including individually, with community groups, and with local organizations. Some central themes of its work are: effective communication about climate change, home energy, transport, food, other kinds of consumption and waste. CCF, started by Rosemary Randall and Andy Brown in 2005, engaged residents of Cambridge in one-to-one conversations about climate change and reducing their carbon footprint. Rosemary Randall's expertise in psychotherapy and Andy Brown's proficiency in low-energy buildings, shaped CCF's approach to carbon reduction as a psychological and technical challenge.

Over the years CCF has run a diverse range of projects aimed at providing both practical advice and psychological support for personal action on climate change. CCF expanded its outreach work in 2008 and became a charity in 2009. The UN climate talks in Copenhagen prompted more political work by CCF. It joined with other groups as a member of the Stop Climate Chaos national coalition, now the Climate Coalition, and remains a member still. Since 2010 it has run Open Eco Homes, where house owners show their low-energy homes to visitors, who feel inspired to make their own home energy savings. In 2012, CCF's work on reducing food footprints blossomed, and it incubated Cambridge Sustainable Food, now a separate organization. Today, it has a growing focus on reducing consumption footprints that involves many other organizations. The importance of psychology in developing climate change action programs is increasingly on the CCF agenda. Rosemary Randall, as mentioned before, has been recognized for her writing and lectures on the psychology of climate change, and for the practical applications she has developed from her research.

[11] For more information see: https://cambridgecarbonfootprint.org/

2 THE IMPORTANCE OF THE CONCEPT OF SUBJECTIVITY: NEW LINES...

Rosemary Randall, has been analysing the rationale for governmental behaviour change policies, offering criticism to this approach, and creating an alternative psychological model for engaging people in change. According to Randall, denial, doubt, grief, anger, confusion, and apathy and are some of the many feelings that spontaneously arise in response to climate change. There is an abundance of information currently available about the importance of changing people's behaviour and, to Randall, this follows the main trend of psychological work on climate change. This work is being done by those with a background in behavioural and cognitive psychology and the focus is on how to take people from a shift in attitude to a shift in behaviour.

Although there is much value in this work, Randall prefers to think in terms of people's *whole selves* and the full complexity of our responses – emotional and spiritual, as well as cognitive and behavioural. To Randall we need to understand ourselves as driven by conflicts – between our desires and our intentions; between our social selves and our spiritual selves; between our needs and our aspirations. In her opinion, this is essentially the human condition. Randall believes that a focus on behaviour alone reduces us to mechanical, controllable beings, who can be manipulated or nudged into the correct responses. However, by looking at the *whole self* – the living, breathing, complex individuals who can love, understand, struggle, and choose – we have a much more complete picture of people in general.

Randall emphasizes the need to understand and answer why climate change is so hard to deal with. In general, she sees the inadequate public response to climate change as largely an information deficit; a difficulty in appreciating dangers that are distant in time or place; or as a tendency for people to seek explanations that confirm their existing cultural assumptions/status quo. Meanwhile, the lack of an adequate political response is usually attributed to lack of political will and leadership; the difficulty of complex negotiations; or the operation of crude nationalism and economic self-interest. Most of these explanations, in Randall's perspective, keep us in the cognitive/rational realm and don't ask the deeper and more uncomfortable questions. Questions about the subjective experience of confronting an issue as life-changing and disturbing as this one or allow us to see the structures of feeling that characterize apparently rational political processes.

In Randall's view psychoanalysis helps us explore the dark underbelly of people's reactions. Its model of the psyche as a place of inner conflict and unconscious, points to the conclusion that all is not as it appears. It allows us to ask the awkward questions that help us connect personal responses to their cultural context, and to the socio-political landscape. It helps us examine questions such as, how do people defend themselves against frightening or distressing news? What lies behind the crude assumptions of 'denial' or 'apathy'? What part do the darker emotions – greed, envy, destructiveness, selfishness – play in maintaining an unsustainable status quo? How do people cope with feelings of insignificance, anxiety, despair, loss, and grief at what might happen? What is hidden by assumptions of rationality, and by the dominant rational discourses of behavioural change and economic self-interest? What are the structures of feeling and defence that characterize the cultural, political, and social systems, the policy decisions, and international negotiations that have failed to deliver, or have opposed action?

In this perspective, to Randall, psychoanalysis suggests that it is important to stop being afraid of our darker feelings. Bring them into the light of shared experience where they can be dealt with creatively. So, it follows that the author explores the experiences of those who try to come to terms with what is called *ecological debt* – which includes processing feelings of shock, disbelief, shame, and guilt. In Randall's interviews she analyses the pain of living in close contact with the knowledge of apparent climate change. She acknowledges the depth of personal change demanded and the need for appropriate support systems.

Psychoanalysis, to Randall, has much to offer to our understanding of the social processes at work in our failure to deal properly with climate change; and to the creation of what she calls the 'safe spaces' that could allow difficult emotional experiences to be confronted and worked through and the political reality to be appropriately engaged with. In her perspective, there is a kind of hope that comes from a stronger connection to reality, an ethical realism that is in the end more sustaining than illusion.

In this chapter we seek to identify and present the relatedness between environmental questions, their diverse impacts on the human psyche, especially in regards to subjectivity. As we have already pointed out, much still needs to be analysed to help face the challenges that environmental changes, including climate change, are bringing to the planet's diverse ecosystems and to the quality of life of human beings, especially to their mental health. Highlighted in this chapter are the actions of many people that are trying to minimize impacts and also help people understand and gain awareness of the severity of this problem. As already stated, this has been our challenge.

References

Adams, M. (2014). Inaction and Environmental Crisis: Narrative, Defence Mechanisms and the Social Organisation of Denial. *Psychoanalysis, Culture & Society, 19*(1), 52–71.

Adams, M. (2017). Trying to Build a Movement Against Ourselves? On Reading Renee Lertzman's Environmental Melancholia: Psychoanalytic Dimensions of Engagement. *Psychoanalysis, Culture & Society, 22*, 220–228. https://doi.org/10.1057/s41282-016-0034-8

Barbosa, S. R. C. S. (1996). *Qualidade de vida e suas metáforas. Uma reflexão socioambiental*. Tese de Doutorado em Ciências Sociais, IFCH, UNICAMP.

Beck, U. (2010). Climate for Change, or How to Create a Green Modernity? *Theory Culture and Society, 27*(2–3), 254–266. https://doi.org/10.1177/0263276409358729

Ben-Asher, S., & Goren, N. (2006). Projective Identification as a Defense Mechanism When Facing the Threat of an Ecological Hazard. *Psychoanalysis, Culture & Society, 11*, 17–35. https://doi.org/10.1057/palgrave.pcs.2100057

Brown, R. S. (2016). Disadapting to the Environmental Crisis. *Psychoanalysis, Culture & Society, 21*(4), 426–433. https://doi.org/10.1057/s41282-016-0004-1

Cardwell, F. S., & Elliott, S. J. (2013). Making the Links: Do We Connect Climate Change with Health? A Qualitative Case Study from Canada. *BMC Public Health, 13*, 208.

Chakrabarty, D. (2017). The Politics of Climate Change Is More Than the Politics of Capitalism. *Theory, Culture & Society, 34*(2–3), 25–37. https://doi.org/10.1177/0263276417690236

Chancer, L., & Andrews, J. (2014). Introduction: The Unhappy Divorce: From Marginalization to Revitalization. In L. Chancer & J. Andrews (Eds.), *The Unhappy Divorce of Sociology and Psychoanalysis* (Studies in the Psychosocial) (pp. 1–14). Palgrave Macmillan.

Cielemecka, O., & Daigle, C. (2019). Posthuman Sustainability: An Ethos for our Anthropocenic Future. *Theory, Culture & Society, 36*(7-8), 67–87. https://doi.org/10.1177/0263276419873710

Clark, N. (2010). Volatile Worlds, Vulnerable Bodies Confronting Abrupt Climate Change. *Theory, Culture & Society, 27*(2–3), 31–53. https://doi.org/10.1177/0263276409356000

Clayton, S. P. D., Wright, P. C., Stern, L., Whitmarsh, A., Carrico, L., Steg, J. S., & Bonnes, M. (2015). Psychological Research and Global Climate Change. *Nature Climate Change, 5*, 640–646. https://doi.org/10.1038/nclimate2622

Correia, D. (2016). Climate Revanchism. *Capitalism Nature Socialism, 27*(1), 1–8. https://doi.org/10.1080/10455752.2016.1140379

Coverdale, J., Richard B., Eugene, V., Beresin, A. M., Brenner Guerrero, A. P. S., Louie, A. K., & Roberts, L. W. (2018). *Climate Change*: A Call to Action for the Psychiatric Profession. *Academic Psychiatry, 42,* 317–323. https://doi.org/10.1007/s40596-018-0885-7

Dunlap, R. E. (2013). Climate Change Skepticism and Denial: An Introduction. *American Behavioral Scientist, 57*(6), 691–698. https://doi.org/10.1177/0002764213477097

Dutta, P., & Chorsiya, V. (2013). Scenario of Climate Change and Human Health in India. *International Journal of Innovative Research & Development, 2*(8), 157–160.

Ebi, K. L. (2013). Is Climate Change Affecting Human Health? *Environmental Research Letters, 8*(3), 31002–31004. https://doi.org/10.1088/1748-9326/8/3/031002

Ebi, K. L., & Semenza, J. C. (2008). Community-Based Adaptation to the Health Impacts of Climate Change. *American Journal of Preventive Medicine, 35*(5), 9749–3797.

Ermert, V., Fink, A. H., & Paeth, H. (2013). The Potential Effects of Climate Change on Malaria Transmission in Africa Using Bias-Corrected Regionalized Climate Projections and a Simple Malaria Seasonality Model. *Climatic Change, 120,* 741–754. https://doi.org/10.1007/s10584-013-0851-z

Euler, J. (2019). The Commons: A Social Form that Allows for Degrowth and Sustainability. *Capitalism Nature Socialism, 30*(2), 158–175. https://doi.org/10.1080/10455752.2018.1449874

Fay, J. (2016). Psychotherapy and Global Transformation. *Psychotherapy and Politics International, 14*(2), 76–83. https://doi.org/10.1002/ppi.1378

Few, R. (2007). Health and Climatic Hazards: Framing Social Research on Vulnerability, Response and Adaption. *Global Environmental Change, 17,* 281–295.

Finley, E. (2019). Beyond the Limits of Nature: A Social-Ecological Perspective on Degrowth as a Political Ideology. *Capitalism Nature Socialism, 30*(2), 244–250. https://doi.org/10.1080/10455752.2018.1499122

Fritze, J. G., Blashki, G. A., Burke, S., & Wiseman, J. (2008). Hope, Despair and Transformation: Climate Change and the Promotion of Mental Health and Wellbeing. *International Journal of Mental Health Systems, 2*(13). https://doi.org/10.1186/1752-4458-2-13

Giddens, A. (2009). *The Politics of Climate Change.* Polity.

Hoggett, P. (2011). Climate Change and the Apocalyptic Imagination. *Psychoanalysis, Culture & Society, 16*(3), 261–275. https://doi.org/10.1057/pcs.2011.1

House, R. (2019). Act Now to Prevent an Environmental Catastrophe. *Psychotherapy and Politics International, 17,* e1484. https://doi.org/10.1002/ppi.1484

Hoggett, P. (2019). Chapter 1. Introduction. In P. Hoggett (Ed.), *Climate Psychology. On Indifference to Disaster* (Studies in the Psychosocial). https://doi.org/10.1007/978-3-030-11741-2_1

Hughes, J. D. (2010). Climate Change: A History of Environmental Knowledge. *Capitalism Nature Socialism*, 21(3), 75–80. https://doi.org/10.1080/10455752.2010.508619

Hulme, M. (2010). Cosmopolitan Climates. Hybridity, Foresight and Meaning. *Theory, Culture & Society*, 27(2–3), 267–276. https://doi.org/10.1177/0263276409358730

IPCC. (2007). *Climate Change 2007: Synthesis Report. Contribution of Working Groups I, II and III to the Fourth Assessment Report of the Intergovernmental Panel on Climate Change*. IPCC.

IPCC. (2014). Summary for Policymakers. In C. B. Field, V. R. Barros, D. J. Dokken, K. J. Mach, M. D. Mastrandrea, T. E. Bilir, M. Chatterjee, K. L. Ebi, Y. O. Estrada, R. C. Genova, B. Girma, E. S. Kissel, A. N. Levy, S. MacCracken, P. R. Mastrandrea, & L. L. White (Eds.), *Climate Change 2014: Impacts, Adaptation, and Vulnerability. Part A: Global and Sectoral Aspects. Contribution of Working Group II to the Fifth Assessment Report of the Intergovernmental Panel on Climate Change* (pp. 1–32). Cambridge University Press.

Josanoff, S. (2010). A New Climate for Society. *Theory, Culture & Society*, 27(2–3), 233–253. https://doi.org/10.1177/0263276409361497

Kabir, M. I., Rahman, M. B., Smith, W., Lusha, M. A., Azim, A., & Milton, A. H. (2016). Knowledge and Perception About Climate Change and Human Health: Findings from a Baseline Survey Among Vulnerable Communities in Bangladesh. *BMC Public Health*, 16(1), 266. https://doi.org/10.1186/s12889-016-2930-3

Lertzman, R. (2012). Researching Psychic Dimensions of Ecological Degradation: Notes from the Field. *Psychoanalysis, Culture & Society*, 17(1), 92–101.

Lertzman, R. (2015). *Environmental Melancholia: Psychoanalytic Dimensions of Engagement* (Psychoanalytic Explorations). Routledge.

Liang, S., Kintziger, K., Reaves, P., & Ryan, S. J. (2017). Climate Change Impacts on Human Health. In E. P. Chassignet, J. W. Jones, V. Misra, & J. Obeysekera (Eds.), *Florida's Climate: Changes, Variations, & Impacts* (pp. 125–152). Florida Climate Institute. https://doi.org/10.17125/fci2017.ch04

Lovelock, J. (2006). *The Revenge of Gaia: Why the Earth Is Fighting Back – And How We Can Still Save Humanity*. Allen Lane.

Marván, M. L., & López-Vázquez, E. (Eds.). (2018). *Preventing Health and Environmental Risks in Latin America*. Springer Nature. https://doi.org/10.1007/978-3-319-73799-7

McCright, A. M., & Dunlap, R. E. (2000). Challenging Global Warming as a Social Problem: An Analysis of the Conservative Movement's Counter-Claims. *Social Problems*, 47, 499–522.

McMichael, A. J., Campbell-Lendrum, D. H., Cordolan, C. F., Ebi, K. L., Githeko, A. K., Scheraga, J. D., & Woodward, A. A, (Eds) (2003). *Climate Change and Human Health Risks and Responses*. : World Health Organization.

Mnguni, P. P. (2010). Anxiety and Defense in Sustainability. *Psychoanalysis, Culture & Society, 15*, 117–135. https://doi.org/10.1057/pcs.2009.33

Parks, B. C., & Roberts, J. T. (2010). Climate Change, Social Theory and Justice. *Theory, Culture & Society, 27*(2–3), 134–166. https://doi.org/10.1177/0263276409359018

Randall, R. (2005). A New Climate for Psychotherapy? *Psychotherapy and Politics International, 3*(3), 165–179. https://doi.org/10.1002/ppi.7103023

Randall, R. (2009). Loss and Climate Change: The Cost of Parallel Narratives. *Ecopsychology, 1*(3). https://doi.org/10.1089/eco.2009.0034

Rekret, P. (2016). A Critique of New Materialism: Ethics and Ontology. *Subjectivity, 9*(3), 225–245.

Rey, F. G. (2017). The Topic of Subjectivity in Psychology: Contradictions, Paths and New Alternatives. *Journal for the Theory of Social Behaviour, 47*(4), 1–20. https://doi.org/10.1111/jtsb.12144

Rey, F. G. (2019). Subjectivity and Discourse: Complementary Topics for a Critical Psychology. *Culture & Psychology, 25*(2), 178–194. https://doi.org/10.1177/1354067X18754338

Rust, M.-J. (2008). Climate on the Couch: Unconscious Processes in Relation to our Environmental Crisis. *Psychotherapy and Politics International, 6*(3), 157–170. https://doi.org/10.1002/ppi.174

Rustin, M. (2013). How Is Climate Change an Issue for Psychoanalysis? In S. Weintrobe (Ed.), *Engaging with Climate Change: Psychoanalytic and Interdisciplinary Perspectives* (pp. 170–185). Routledge.

Samuels, B. (2015). A Psychoanalytic Intervention to Fight Climate Change: Reading This Changes Everything. *Psychoanalysis, Culture & Society, 20*, 86–89. https://doi.org/10.1057/pcs.2015.5

Segal, H. (1987). Silence Is the Real Crime. *International Review of Psychoanalysis, 14*(3), 3–11. Reprinted with postscript 'Perestroika, The Gulf War and 11 September 2002', In Covington, C., Williams, P., Arundale, J. and Knox, J. (eds), *Terrorism and War: Unconscious Dimensions of Nuclear Violence* (pp. 249–262). Karnac.

Seixas, S. R. C. (2008). Transformações socioculturais contemporâneas e algumas implicações nos diagnósticos na área de saúde mental. *Mudanças – Psicologia da Saúde, 16*(1), 1–9. https://doi.org/10.15603/2176-1019/mud.v16n1p1-9

Seixas, S. R. C., & Ferreira, L. C. (2020). Mudanças climáticas e Covid 19: perspectivas futuras para enfrentamentos de eventos extremos, *Jornal da UNICAMP*, 05 out 2020. Available: https://www.unicamp.br/unicamp/ju/artigos/ambiente-e-sociedade/mudancas-climaticas-e-covid-19-perspectivas-futuras-para

Seixas, S. R. C., & Nunes, R. J. (2017). Subjectivity in a Context of Environmental Change: Opening New Dialogues in Mental Health Research. *Subjectivity, 10*, 294–312. https://doi.org/10.1057/s41286-017-0032-z

Seixas, S. R. C., Hoeffel, J. L. M., Bianchi, M., & Santos, A. (2010). Qualidade de vida, ambiente e subjetividade na APA Cantareira. In Hoeffel, J. L. M., Fadini, A. A. B., & Seixas, S. R. C. (Orgs.) *Sustentabilidade, Qualidade de Vida e Identidade Local Olhares Sobre as APAs Cantareira, SP e Fernão Dias (MG)* (pp. 115–134). RiMa.

Seixas, S. R. C., Renk, M., Hoeffel, J. L. M., Conceição, A. L., & Asmus, G. F. (2012). Global Environmental Changes and Impacts on Fishing Activities in the Northern Coast of Sao Paulo, Brazil. Urban Areas and Global Climate Change. *Research in Urban Sociology, 12*, 299–317. https://doi.org/10.1108/S1047-0042(2012)0000012015

Seixas, S. R. C., Hoeffel, J. L. M., Botterill, T. D., Vianna, P. V. C., & Renk, M. (2014). Violence, Tourism, Crime and Subjective Mental Health: Opening New Lines of Research. In H. Andrews (Ed.), *Tourism and Violence* (pp. 145–164). Ashgate.

Seixas, S. R. C., Hoeffel, J. L. M., Renk, M., Asmus, G. F., & Lima, F. B. (2016). Weather Variability and Climate Change Impacts on the Mental Health of a Seaside Community. *Journal of Scientific Research & Reports, 11*(3), 1–17.

Seppänen, J. (2011). Lost at Sea: The Freudian Uncanny and Representing Ecological Degradation. *Psychoanalysis, Culture & Society, 16*, 196–208. https://doi.org/10.1057/pcs.2010.32

Shove, E. (2010). Social Theory and Climate Change Questions Often, Sometimes and Not Yet Asked. *Theory, Culture & Society, 27*(2–3), 277–288. https://doi.org/10.1177/0263276410361498

Skrimshire, S. (2019). Deep Time and Secular Time: A Critique of the Environmental 'Long View'. *Theory, Culture & Society, 36*(1), 63–81. https://doi.org/10.1177/0263276418777307

Swyngedouw, E. (2010). Apocalypse Forever? Post-political Populism and the Spectre of Climate Change. *Theory, Culture & Society, 27*(2–3), 213–232. https://doi.org/10.1177/0263276409358728

Szerszynski, B. (2000). On Knowing What to Do: Environmentalism and the Modern Problematic. In S. Lash, B. Szerszynski, & B. Wynne (Eds.), *Risk, Environment & Modernity* (pp. 104–137). SAGE.

Szerszynski, B., & Urry, J. (2010). Changing Climates: Introduction. *Theory Culture & Society, 27*(2–3), 1–8. https://doi.org/10.1177/0263276409362091

Tschakert, P., Barnett, J., Ellis, N., Lawrence, C., Tuana, N., New, M., Elrick-Barr, C., Pandit, R., & Pannell, D. (2017). Climate Change and Loss, as if People Mattered: Values, Places, and Experiences. *Climate Change, 8*(5), e476. https://doi.org/10.1002/wcc.476

UN Environment (Ed.). (2019). *Global Environment Outlook – GEO-6: Healthy Planet, Healthy People.* Cambridge University Press. https://doi.org/10.1017/9781108627146

United Nation Sustainable Development (UNSD) (1992). *United Nations Conference on Environment & Development – Agenda 21*. Rio de Janeiro, Brazil, 3 to 14 June 1992. Available: sustainabledevelopment.un.org/content/documents/5987our-common-future.pdf

United Nations (UN) (1968). *The Problems of Human Environment*. General Assembly A/7291 – 23 Session. New York.

United Nations (UN) (1972). *United Nations Conference on the Environment*, 5–16 June 1972, Stockholm – Stockholm Declaration and Action Plan for the Human Environment and Several Resolutions, Report of the United Nations Conference on the Human Environment. New York. undocs.org/en/A/CONF.48/14/Rev.1

United Nations (UN) (2005). *Report of the World Social Situation 2005: The Inequality Predicament*. New York. Available: un.org/esa/socdev/rwss/docs/2005/rwss05.pdf

United Nations Environment Programme (UNEP). (2002). *Annual Report*. United Nations (UN).

Van Susteren, L. (2018). The Psychological Impacts of the Climate Crisis: A Call to Action (Editorial). *BJPsych International, 15*(2), 25–26. https://doi.org/10.1192/bji.2017.40

Wei-Lun, T., McHale, M. R., Jennings, V., Marquet, O., Hipp, J. A., Leung, Y., & Floyd, M. F. (2018). Relationships Between Characteristics of Urban Green Land Cover and Mental Health in U.S. Metropolitan Areas. *International Journal of Environmental Research Public Health, 15*, 340. https://doi.org/10.3390/ijerph15020340

Weintrobe, S. (2013). Chapter 1: Introduction. In S. Weintrobe (Ed.), *Engaging with Climate Change. Psychoanalytic and Interdisciplinary Perspectives* (pp. 1–15). Routledge.

IMPORTANT LINKS

https://climateoutreach.org/
https://climateoutreach.org/climate-communication-awards/#
https://climateoutreach.org/we-analysed-the-polls-this-is-what-the-uk-thinks-of-the-climate-crisis/
https://sustainabledevelopment.un.org/outcomedocuments/agenda21
https://www.cardiff.ac.uk/psychology/research/social-and-environmental/centre-for-climate-change-and-social-transformations-cast
https://www.cardiff.ac.uk/sustainable-places
https://www.cardiff.ac.uk/water-research-institute

CHAPTER 3

Building New Perspectives and Approaches to Our Common Future on Climate Change and Subjectivity: Agenda 2030 and Human Rights

Abstract This chapter shows that the idea of a common future is highlighted through awareness that a global collapse may impact the entire planet. Living with a pandemic like Covid-19, since March 2020, has shown that this is a concrete fact and what it means for the planet. For this reason, the choice and recognition of the Global Agendas, Universal Declaration of Human Rights (UN, *Universal Declaration of Human Rights (UDHR)*. Available: un-org/en/about-us/universal-declaration-of-human-rights, 1948), and Agenda 2030 (UN, *Transforming our World: The Agenda 2030 for Sustainable Development*. Available: un.org/ga/search/view_doc.asp?symbol=A/RES/70/1&Lang=E, 2015) are concrete options that lead us to recognize that we do have a Common Future. Our predatory way of acting, and the absence of criticism of this model of development, need to be rethought whether through a radical critique of this model or by building new values and ways of living, or *the good life*.

Keywords Our common future · Subjectivity · Climate change · Agenda 2030 · Human rights

The absence of a common world to share is driving us mad. (Latour, 2020: 10[1])

Learning about subjectivity and climate change has revealed the complex interconnected relationship between three things. Figure 3.1 highlights our vision of building new perspectives to deal with the challenges contemporary society is facing.

The first thing we need to recognize is that we have a common future, and, as the Brundtland Report has already pointed out (UN, 1987),[2] this common future allows us to deal with our common global scenarios and challenges. It is important to emphasize, as we illustrate in Fig. 3.1, we have been affected by and have been subjected to a global collapse – especially in the last decades. The Covid 19 pandemic is evidence of our limitations as national and subnational populations.

Jappe et al. (2020: 33), calls our attention to something important. He tells us that "*epidemics are histories great characters*". This has been true since the early Neolithic and Agricultural Societies and is still true of our Capitalist Society of today. The authors highlight that the Covid 19 pandemic is no different than earlier epidemics in that it brought to light the

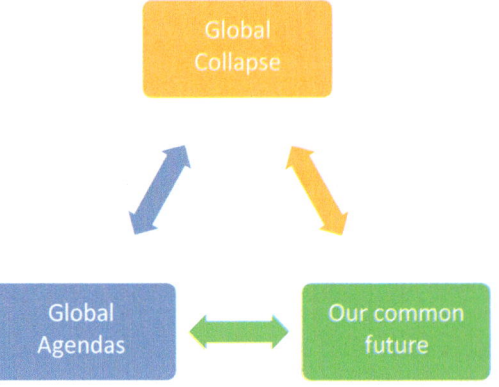

Fig. 3.1 Common Future: Inter-social environmental relationships and their challenges. (Source: The authors, 2021)

---

[1] Latour, B (2020). Onde aterrar? – Como se orientar no Antropoceno. Rio de Janeiro: Bazar do Tempo.
[2] United Nations (1987). Report of the World Commission on Environment and Development – Our Common Future (also known as the Brundtland Report, in reference to the Report's coordinator and Prime Minister of Norway at the time, who is also a leading feminist, physician and advocate for sustainable development).

limitations between the social and the biological. It poses the same essential questions about its causes, how it spread, and its social impact. The pandemic continues to be a product of capitalist social life.

We opted to treat this as global collapse and not just a crisis because this great global disaster is the result of limitations of the exclusive development of a capitalist system that has made most of the world's population vulnerable and put them at a disadvantage. Our hypothesis is based on an idea defended by many researchers. The authors emphasis that the term *social natural catastrophe*, that is already being used for environmental devastations resulting from climate change, is directly applicable to the global situations that we are now facing. Independent of the cause of the spread of Covid 19 on the planet, that will still require many global studies to elucidate, it is important to underscore that Sars-Cov-2 was the detonator, and not the bomb that worsened a structural crisis that we have been facing for a while now (Jappe et al., 2020: 34).

At play in this logic is wondering what it means to have a developmental model, aggravated by a pandemic, that brings about global collapse. We're looking to offer real alternatives for facing this paradox. Besides questioning and provoking a critique, we offer reflection on what we believe could be one possible scenario for facing these challenges.

What can be critiqued about the developmental model currently in vogue? The first thing we'd like to point out is the meaning of development and what is understood about this concept. Lang (2016) pointed out that development is a natural process where plants and animals and humans develop from a seed, an egg, an embryo until they reach maturity, explaining a linear and unlimited process. This way of looking at the concept is present in many approaches and always seen as something positive. From the point of view of dynamic systems of evolution, reaching quality of life and good living conditions for a population, is more limited because it is impossible for the entire planet to have the same model of development. This is due to differences in economics, politics, geographies, and natural resources, or as Acosta (2016a) says, in function of the politics of extractivism native to rich counties in their relations with poor countries.

Lang (2016) suggests that the concept of development must stop being seen as something positive, because it is a deceptive promise. The great majority of the populations of the geo political south, the so called 'cooperation for development' have transferred way more of the South's resources to the North than vice versa. This alone makes this concept a fallacy. The 'donor economies', (the global North), export technology to poor countries and in doing so guarantee employment and income for their own economies, and not the contrary. This was evident in research done in 2014 on global financial flows that showed that, in the existing system, for every US 1 million dollars that enters a developing country, this same country loses more than US 2 million dollars (Lang, 2016: 27, note 2). For seven decades this commitment has not been fulfilled. For seven decades the industrialized Global North has been trying to impose a model that would permit a way of life similar to their own. This mode of life is characterized by unlimited consumption, and individually self-managed careers, seen as the essence of quality of life. This has not, and will not, be achieved by maintaining this model in the South.

The author explores another aspect that is directly related to the inequalities inherent in the model of the Global North. These inequalities have provoked loneliness, anguish, depression, permanent stress as an underlying cause of physical and mental illness. The inequalities have instigated a lack of coexistence, sharing of collective projects, debt, and exacerbated patterns of consumption – many of which aren't even necessary to wellbeing. It is already well known that a real correlation between GDP growth and a population's happiness is non-existent. Recent data shows that the countries with the most money in circulation for consumption (Denmark, Norway, Australia and the United States) are also the countries with the highest rates of suicide. The same holds true for Germany where despite a robust economy with positive macroeconomic indicators, the inequality between rich and poor has substantially increased, and shows that 1 in 5 children live in poverty (Lang, 2016: 28, note 5). It's devastating to consider the objective conditions of a country like this.

The biggest criticism the author makes in his article, gives us another real way of building something which is an alternative to development, not an alternative form of development. Sustainable development is a good example. It recognizes and includes the knowledge and multiple world views of those involved by listening to them, then finding ways of constructing projects contextualized in collective learning and the vocation of the regions where they are to be implanted. Tailoring the alternative to a specific population will supplant not only a capitalist-based work-capital relationship but also overcome colonialism, patriarchy and the predatory relationship with nature that this system was built on. Other fundamental aspects to consider in an *alternative to development* project are the transformations in subjectivity and in interpersonal relationships in the practices of daily living. Also, the incorporation of various academic disciplines and other current lines of thought – such as ecology, feminism and the actors fundamental to this process; first nations, women, LGBTQI+ and others – in order that an alternative project may truly be thought out and consolidated (Lang, 2016: 44).

This brief critique of the concept of development is contingent upon opening to building new alternatives that consider creating situations more adequate for a common future. A common future where all of the aspects mentioned are included, and where the immense challenges that we will have to face from global collapse (due to health crisis, climate emergency, and the crisis of biodiversity) can be supplanted as a result of other viewpoints and reflections.

It is important to recognize that the need to consolidate collective action strategies to confront these challenges, requires tools that allow us to think of different ways and new approaches for this construction. For this reason, the choice and recognition of the Global Agendas, Universal Declaration of Human Rights (UN, 1948), and Agenda 2030 (UN, 2015) are concrete options that lead us to recognize that we do have a Common Future. Our predatory way of acting, and the absence of criticism of this model of development, need to be rethought whether through a radical critique of this model or by building new values and ways of living, or *the good life*.

The idea of a common future is highlighted through awareness that global collapse impacts the entire planet. Living with a pandemic like Covid-19, since March 2020, has shown that this is a concrete fact and what it means for the planet. The three circumstances shown in Fig. 3.1 present a way of understanding this. We will discuss this and actions that can be taken.

Reflection on quality of life, environmental changes, and subjectivity has been the focus of our research and analysis for some decades. We have taken many paths in our work during this time. In our approach, we defend that it is impossible to think about macro concepts without taking into consideration the concept of subjectivity; and that subjectivity is part of a wider reflection that can be deepened through the eyes and framework of psychoanalysis.

As we have already discussed, psychoanalysis, through its theoretical views and practical motivations, has contributed to the understanding of environmental change, ecology, the degradation of nature, and the social environmental impacts brought about by climate change. This represents a huge advancement. Until the nineteenth century this science was built on looking exclusively at the individual, but what we have learned since is that the barrier of 'within' and 'without' doesn't make sense today. Perhaps it never did make sense. This is because what happens subjectively in the outer world, has an impact within the individual and on subjectivity. But how can we broaden this perspective to get beyond subjective suffering and build connections and networks of collective action to face these challenges? To answer this question, we based our argument on teachings we have received from a number of authors. These authors have made some powerful contributions that underly our reflection from another perspective.

We recognize that humanity has *A Common Future* and that we are being submitted to a powerful global collapse. We also recognize that this common future can only be contemplated if we consider both civilizing agendas – The Universal Declaration of Human Rights and Agenda 2030. To this end, we will bring back key authors to complete this picture.

Acosta (2016b),[3] in her inspiring work, supposes that *The Good Life* is an opportunity to collectively build a new way of life. At the same time, we might ask ourselves if it's possible to implement a new social order within Capitalism. (Acosta, 2016b: 25). For this author, this social order is tied to what we believe and to what she calls attention to in her work as Human Rights and Rights of Nature, which we pointed out are the Global Civilizational Agendas.

Another important point she made is that it's no use only putting the *Good Life* into a country's constitution, as this alone can't overcome the difficulties within a system that, in essence, is a civilization of inequality and devastation. In the same way the civilizing agendas can't be merely words. These agendas need to be seen as something dynamic that mobilizes us as citizens and subjective beings. We need to engrain them in our personal, collective and political actions. We need to find empathy, new ways to act, and new forms of coexisting that are subjective learnings. Or as Acosta highlights:

> People should organize themselves to recuperate and assume control of their own lives. However, this isn't only about defending the work force or getting time off for workers – in other words this isn't just a matter of opposing the exploitation of workers. What's also at play is the defence of

---

[3] Acosta, A. (2016b). O bem viver: uma oportunidade para imaginar outros mundos. São Paulo: Autonomia Literária, Elefante, 264p.

life against anthropocentric schemes of productive organization causing the destruction of the planet. (Acosta, 2016b: 27)

According to Acosta (2016b), what's at play in this developmental model is the need to defend life against a model that is primarily an anthropogenic scheme of production. This form of organization, as we have already presented, is what is destroying the planet and is responsible for climate change, for the accumulation of riches for the few, and for increased world social inequality.

The author points out that it is fundamental to overcome inequality and that inequality will only be supplanted through decolonization and the elimination of power centred in patriarchy. This must be in alliance with overcoming racism, all forms of oppression against women, and LGBTQI+ individuals, which is still rampant in our societies. It is fundamental to overcome these barriers to be able to construct the dignity of life. This effort shouldn't rely on purely technical solutions. Solutions should also be recreated out of collective community attitudes of solidarity and empathy.

Through the *Good Life*, Acosta's line of thought offers a way to question the unviability of our life style. The author points out that limitless material growth could end in collective suicide because endless growth is based on inexhaustible use of resources and a market capable of absorbing everything that is produced (Acosta, 2016b: 34). It has already shown itself incapable of promoting development in the way we have criticized. Data on the distribution of the world's richest shows the immense inequality that we face. The second report from Oxfam Brazil (2020) showed that 2.153% of the world's billionaires hold more wealth than the 4.6 billion people – or around 60% of the world's population. Global inequality is record high and the number of billionaires has doubled in the last decade. Oxfam also highlighted that Brazil is one of the most sexist economies in the world. We see millions of women and girls spending a good part of their lives doing domestic work and doing unpaid caregiving without access to public services that could help them with these important duties. Women in the world do more than 75% of these unpaid jobs. Because of this, they frequently work less hours at their jobs or have to abandon them because of the amount of hours spent on caregiving. Throughout the world, 42% of women weren't able to get work because they are responsible for all of the domestic work. This situation applies to just 6% of men.

What appears to be a contradiction, or an enormous paradox of our time, is that it was believed that the surprising advance in science and technology would open an immense field of possibilities for humanity. The dilemma is that not everyone has benefitted or will benefit from these conquests. A great part of humanity does not have equal access to computers and communication technologies, or access to even basic services like environmental sanitation, potable water, education equipment, and health.

Shockingly, 2020 revealed when Covid-19 called for social isolation there were not enough resources to serve the entire planet. Many societies lacked efficient sanitation conditions, water, electricity, internet, and the food security needed for everyone. This intensified the health crisis inexorably for the whole planet and brought death to more than 4,027,000 people by June of 2021. The United States, Brazil and India make up nearly 1,500,000 of these deaths (WHO, 2021). The hard realization has been that,

1. not everyone has access to basic goods and,
2. the first people hit were the most vulnerable and couldn't maintain even a minimum of health security when faced with such a devasting pandemic. They needed to keep their jobs, usually the most precarious, to keep from dying of hunger.

Adiche (2019: 26), alerted that we must escape consolidating to one unique story, because, one unique story creates innumerable stereotypes. Stereotypes aren't necessarily lies, but are incomplete, and become the only narrative. This idea applies to the capitalistic development claim that any proposal will be positive for everyone. That turns out to be a fallacy. This is evident because climate change is global in scope.

This was highlighted by Ferreira et al. (2020) when they stated that it is fundamental to realize that climate change permeates all countries on the planet and that national and local governments are important and interested parties for building ways to respond to the risks of climate change. These governments can and should create laws and consolidate institutions to deal with the many risks on different scales. At the same time, they should consider the fundamental role of international climate politics and that subnational and national governments are important. On the other hand, this involves a wide spectrum of diverse social actors ranging from government agents to civil groups and individual consumers.

This shows the importance of understanding this problem beyond one theoretical and historical narrative. All of the agents and actors have a fundamental role in facing climate change. Consequently, the subjectivity of all of these actors is part of the analysis of this history.

In this way, our contribution to this reflection in no way discards any of these aspects. Instead, it reinforces that if we have not learned to deeply question a developmental model that relies exclusively on degradation, mass consumption, and profit, and continue to believe this model is the only way possible for economic growth, we will not achieve a new way of facing the perverse dynamics of climate change and social environmental degradation. And, we will not be able to create security for the population of the planet. It is impossible to consider a global mitigation plan, population adaptation, maintain ecosystems and conserve biodiversity without understanding that this is a global action. It involves all national dimensions and all of the different social and political actors in these contexts.

Because of this, we support a complex approach like the one seen in Fig. 3.1 and avoid a single historical outlook. In this approach three things are considered: Global Collapse – Our Common Future – Global Agendas. These three aspects include awareness that we are facing a global collapse due to a failed developmental project for the majority of the world's population, that we share a common future, and that the global agendas we have mentioned can help by offering other alternatives for achieving an alternative developmental model. This model includes aspects directly related to the dignity of life.

What can we learn from these two important agendas in order to overcome a single historical narrative, consider subjectivity, and build a new model of society? If we pay attention to the Universal Declaration of Human Rights (UDHR), as well as to Agenda 2030 we clearly see that, though they are presented in different contexts, they both bring up these approaches. Agenda 2030 (UN, 2015; IPEA, 2018) is understood as a plan of action for people, the planet and for prosperity. It also seeks to strengthen universal peace and greater freedom. The 17 objectives have to do with 5 basic themes:

- *People* – End poverty and hunger in all of its forms and dimensions and guarantee that all human beings can reach their potential in terms of dignity and equality within a healthy environment.
- *Planet* – Protect the planet from degradation by changing habits of consumption, creating sustainable production, sustainable

management of natural resources, and urgent measures to combat climate change in order to care for the needs of present and future generations.
- *Prosperity* – Ensure all human beings have prosperity and personal fulfilment, and that economic, social and technological progress occurs in harmony with nature.
- *Peace* – Promote peaceful, just and inclusive societies free from fear and violence. There can be no sustainable development without peace and there can be no peace without sustainable development.
- *Partnership* – Through establishing a revitalized Global Partnership for sustainable development based on a spirit of strengthened global solidarity concentrated on the needs of the poorest and most vulnerable and with participation of all countries, interested groups and people.

From the perspective of these 5 basic themes, the 17 Objectives for Sustainable Development involve the following proposals:

Objective 1. To end poverty in all forms, everywhere;
Objective 2. To end hunger, achieve food security, improved nutrition and sustainable agriculture;
Objective 3. To insure health and wellbeing for all people of all ages;
Objective 4. To insure inclusive, equitable and quality education throughout life for everyone;
Objective 5. To achieve gender equality and equal training opportunities for all women and girls;
Objective 6. To insure sustainable management and access to water and basic sanitation for everyone;
Objective 7. To insure reliable, sustainable, modern and affordable energy for everyone;
Objective 8. To promote sustainable and inclusive economic growth, full and productive employment and decent work for everyone;
Objective 9. To build resilient infrastructure, promote inclusivity, create sustainable industrialization, and foment innovation;
Objective 10. To reduce inequality in and between countries;
Objective 11. To make human settlements inclusive, resilient and sustainable;
Objective 12. To insure sustainable means of production;

Objective 13. To take urgent measures to combat climate change and it's impacts;

Objective 14. To conserve the sustainable use of oceans, seas, and marine resources for sustainable development;

Objective 15. To protect, recuperate and promote the sustainable use of terrestrial ecosystems, sustainably managed forests, combat desertification, stop and reverse the loss of biodiversity;

Objective 16. To promote peaceful and inclusive societies for sustainable development, provide access to justice for everyone, and build efficient, responsible and inclusive institutions on all levels;

Objective 17. To strengthen ways to implement and to revitalize global partnership for sustainable development.

The relationship of these objectives with the UDHR can be summarized from paragraph 1 of Article XXV:

> Every human being has the right to a standard of living that insures they and their family have health, wellbeing, food, clothing, shelter, medical care, essential social services, the right of insurance in case of unemployment, sickness, disability, loss of spouse, old age and other unforeseen losses that threaten subsistence. (UN, UDHR, 1948)

In our understanding this is the synthesis of wellbeing and dignity of life. It is also an individual and collective pattern of life that will only be achieved if we consider the eradication of poverty, the search for equity, the fight against gender injustice, the defence of quality health and education for all, and the fundamental rights of identity and inclusion as laid out in Agenda 2030.

In order for this we need to find a way for these two agendas to intersect and get beyond the idea that the whole world is a market and everything is merchandise, Krenak (2020: 45), and that everything outside of ourselves is merchandise. We must disregard the idea that political power and its choices can create temporary secure spaces for communities, that however devoid of any true meaning of collective sharing they are, make us believe that we need to exhaust the forests, water resources, and other living species as the only way human communities can survive. This should not be our reality and even less should it be our premise (Krenak, 2020).

In this chapter we described the core of our vision in constructing this book. We recognize three dynamic and interconnected aspects that share a common future with both its positive and negative aspects. Positive to the extent that we represent a great network of connections that is the result of technological advances in communication. Also, to be taken into account is our ability to be in different parts of the planet far more quickly and the increase in consumption that is the result of the processes of globalization. On the negative side, we are aware that these positive aspects aren't equal for everyone and that we have had a significant increase in poverty around the world. Also, what can reach one region reach others, and in a very short period of time, reach everyone. We can't forget that these negative facts were cruelly confirmed in the spread of the Covid-19 pandemic. At the same time, the perception of global collapse is part of an important analysis of a dynamic that is made up of a of health crisis of enormous proportions, a climate emergency, and a biodiversity crisis. In our conception, this collapse is the result of a model of development that is in essence predatory and exclusionary. And finally, we propose that the way out should be based on the global agendas because they can help us find ways of overcoming or mitigating a collapse of this size.

These global agendas allow us to promote a new form of subjective construction where not only the individual and their most intimate aspects can be prioritized, but a new set of social values that include empathy and solidarity can be present. It is in this sense that psychoanalysis contributes to effective understanding and contribution to the subjective actions of humanity.

This being said, we should consider that the enormous size of the challenges of climate change, extreme events, pandemics, and globalization of social environmental problems we face, and will continue to face can only be treated or mitigated through a strong process of adaptation if we recognize our collective role, and if we have awareness that as a society, we have a Common Future. Only through the Global Agendas will we be able to meet these challenges and not fall into the barbarity that is the fruit of social inequality and the immense political and social obstacles that we will have gone forward. This perspective emerges as a new theoretical and methodological consolidation to contribute to this conversation and the future challenges of the contemporary world. It is also a means of surpassing this seemingly dystopian moment by returning to the search for the great social utopias that have fed generations.

References

Acosta, A. (2016a). Extrativismo e neoextrativismo. Duas faces da mesma maldição. In G. Dilger, M. Lang, & J. Pereira Filho (Orgs.). *Descolonizar o Imaginário: debates sobre pós-extrativismo e alternativas ao desenvolvimento* (pp. 47–85). Fundação Rosa de Luxemburgo/Ed. Elefante.
Acosta, A. (2016b). *O bem viver: uma oportunidade para imaginar outros mundos* (p. 264). Autonomia Literária, Elefante.
Adiche, C. N. (2019). *O Perigo de uma história unica*. Companhia das Letras.
Ferreira, L. C., Barbi, F., & Barbieri, M. D. (Org.). (2020). *Dimensões humanas das Mudanças Climáticas no Sul Global*. Editora CRV/FAPESP.
Instituto de Pesquisa Econômica Aplicada (IPEA). (2018). *Agenda 2030. ODS – Metas Nacionais dos Objetivos de Desenvolvimento Sustentável*. IPEA.
Jappe, A., Aumercier, S., Homs, C., & Zacarias, G. (2020). *Capitalismo em quarentena. Notas sobre a crise global*. Elefante.
Krenak, A. (2020). *Ideias para adiar o fim do mundo*. Companhia das Letras.
Lang, M. (2016). Introdução. Alternativas ao desenvolvimento. In: G. Dilger, M. Lang, & J. Pereira Filho (Orgs.), *Descolonizar o Imaginário: debates sobre pós-extrativismo e alternativas ao desenvolvimento* (pp. 26–44). Fundação Rosa de Luxemburgo/Ed. Elefante.
Latour, B. (2020). *Onde aterrar? Como se orientar no Antropoceno*. Bazar do Tempo.
Oxfam Brasil. (2020). *Bilionários do mundo têm mais riqueza do que 60% da população mundial*. Oxfam Brasil. Access: 16.07.2021.
United Nations (UN). (1948). *Universal Declaration of Human Rights (UDHR)*. Available: un-org/en/about-us/universal-declaration-of-human-rights
United Nations (UN). (1987). *Report of the World Commission on Environment and Development: Our Common Future*. Oxford University Press.
United Nations (UN). (2015). *Transforming Our World: The Agenda 2030 for Sustainable Development*. Available: un.org/ga/search/view_doc. asp?symbol=A/RES/70/1&Lang=E
World Health Organization (WHO). (2021). *COVID-19 Weekly Epidemiological Update*. Edition 48, Published 13 July 2021. WHO.

CHAPTER 4

Conclusion

Abstract In this book, we presented a new approach that allows us to continue with the reflection on quality of life, subjectivity and climate change, and also we tried to contribute to a new approach to establish a deeper reading of the meaning of climate change and the role of new theoretical, interdisciplinary approaches, to present another look on the subject.

Keywords A new approach • Global Agendas • Climate change

> If Covid-19 is a spectacular expression of a planetary impasse for humanity, it means nothing more or less than restoring a habitable earth where she can offer a breathable world to everyone. Will we be capable of rediscovering that we belong to the same species and our unbreakable connection to the whole of life? Maybe this is the true question we need to answer, before we shut the door once and for all. (Mbembe, 2020[1])

---

[1] Achille Mbembe, philosopher, historian and professor of history and political science at the University of the Witwatersrand (South Africa) and Duke University (USA), in the article "O direito universal à respiração", published by AOC media – Analyze Opinion Critique https://aoc.media/ and reproduced by Buala and Carta Maior, 14-04-2020, with translation by Mariana Pinto dos Santos and Marta Lança, and available at: http://www.ihu.unisinos.br/78 news/598111 the universal right to breath article de achille mbembe

When we created this book, our ambition was to bring to light the importance of *subjectivity* in order to understand environmental change. Years of research pushed us to recognize the dynamic nature and the importance of this concept. Basically, we believe that if the individual is not paying attention, and able to recognize the impacts of these changes on their subjective world, and at the same time recognize the importance of their individual responsibility in building collective action, we will be unable to reach any substantial goals.

We have been studying environmental changes, especially climate change, from the perspective of the human dimension for long enough to be able to include Global Agendas – UDHR and Agenda 2030. We have recognized the concrete possibility that they can play a fundamental role in advancing this conversation and most especially the importance they place in defending the dignity and quality of life. We must warn, however, that in a significant number of places on the planet, especially in the global south, regions and countries have been plagued by great social inequality, gender inequality, prejudice, racism, xenophobia and degradation of natural resources. Also, since conservatives and the extreme far right have come to power, many of the hard-won social and democratic accords and social policies are being dismantled, one by one. We see this particularly in Brazil, post 2018.

However, not even in our greatest nightmares could we imagine that the year 2020 would bring the cruellest example of a biologically, socially, and environmentally extreme event that humanity has ever experienced. The Covid-19 pandemic is a good example of what humanity may face from now on from the climate emergency. As scientists that observe environmental change, in a holistic and interconnected way, we recognize the significance of the pandemic and felt it was imperative to address this subject in our book. This book has been impregnated by the coincidence of being written at the same time as this historic, social moment. What we have written here draws from the well of analyses that very well may contribute to future scenarios. Even though we were often unable to share Mbembe's (2020) hope, we nonetheless chose him to begin our final reflections and remain open to the call to "*restore a habitable Earth*" (Mbembe, 2020).

In conclusion, it is essential to bring up the most important points used to analyse the fundamental problems that were examined. We nurtured the idea of subjectivity and its different approaches and contributions to the environmental and climate change conversations, in order to promote a better quality of life and to defend the dignity of life. We prioritized the importance of the interdisciplinary, and sought to demonstrate the role of psychoanalysis for understanding the concept of subjectivity, and how this understanding can help construct a positive and effective response to climate change. We used this same interdisciplinary approach to address environmental questions and impacts related to the subjective and the human psyche. At the same time, we considered the importance of the role of psychoanalysis not only from a theoretical standpoint but from a practical standpoint as well. In addition, we looked at what constitutes support networks, not only for individuals but for building collective actions. These support networks spread scientific knowledge and educate professionals who can, in turn, create new approaches through subjectivity to help face climate change. A new approach, using what psychoanalysis can offer, will have a strong influence on these studies by taking into account the impact on human subjectivity.

By reading the theoretical literature and publications from a group of mental health professionals on climate change, we perceived that this contemporary subject finds itself in an immensely challenging situation. On one side, it is possible to see that, since at least the first decade of the twenty-first century, human activity has played a crucial role in environmental degradation, the planetary loss of biodiversity, and in creating climate change. On the other side, we recognize that there is also a strong impact on human subjectivity. With this in mind we have recognized and identified three approaches that effect human beings and society as whole:

1. The first is the belief that scientific discoveries and human beings play a collaborative role in climate change and that there is an urgent need for collective action.
2. Another position is that climate change is occurring but human actions play *no* part in it and that it is a natural, geological process.
3. A third position comes from a group that denies and justifies the existence of climate change. This has resulted in the lack of investment and paradigm shift needed to face these challenges.

While considering these three aspects and their impact on human subjectivity, we perceive that denial is an intense defence mechanism. Denial of the role of science, and the available scientific data, is an exhaustive contradiction. Especially when you consider the reality of extreme events like flooding, droughts, etc. that are already linked to climate change.

It is also important to recognize the role of science to advance social, cultural and environmental development. Also, to be noted is the role of anthropogenic actions in recent environmental changes. The big leap was seeing the role science plays, while at the same time investing in the approach synthesized in Fig. 3.1. This figure illustrates three points that have guided our understanding that we have a Common Future and that we are undergoing Global Collapse:

- From a health crisis as a result of the Covid-19 pandemic,
- From a climate emergency, which has been forming for decades; and from the accelerated degradation of biodiversity, that has reached the entire planet with the most vulnerable groups facing the worst consequences.
- Lastly, we realize the importance of including the two Global Agendas: Universal Declaration of Human Rights (UDHR) and Agenda 2030. These agendas are concrete ways of building climate politics, that recognize the responsibility of humanity. The agendas can be universally managed and, at the same time, hold both national governments, subnational governments, and other social actors accountable. Through these agendas, individuals can revaluate their values, behaviour, and their role in global governance.

These aspects make it possible to build strong sustainability, and change behaviour and social values in such a way that new ways of behaviour, empathy, and living together are learned subjectively. Building the *Good Life* is a concrete alternative to the predominant capitalist developmental model. This brings awareness of both Global Collapse and of a Common Future. This awareness allows us to overcome the idea that the world is merchandise and that social and environmental exploitation are the only strategies for generating value and wealth.

This stance is based on strong political bias and is the fruit of conservative and developmental discussions. Its emotionally constructed aspects, with their large subjective impact, need to be overcome. This position involves attempts to delay implementation of the various changes already proposed. It also keeps mechanisms and structures in place. This brings comfort and security to certain social groups by maintaining consolidated economic and production systems in place. It is also a way to avoid feelings and emotions we often don't want to face. These positions also reduce the need for accelerated change and lead us into situations that would not be considered quality of life.

Even though there is a real need for change, this doesn't always happen as quickly or dynamically as we desire. We have observed that measures previously implemented are often insufficient. Nonetheless, it is again relevant to mention that various human activities are directly related to climate change. Also, countless human activities are increasingly affected, and this has significant impacts on the quality of everyday life.

Innumerable risks are sited in analyses on climate change and sustainability. To mitigate and resolve these situations requires diverse actions. Some of these are in the process of being implemented, and others are continuously delayed. Their effects and solutions are considered objectively and subjectively complex and uncertain.

Our approach attempts to find a new line of research and understanding to help the present problem of climate change. Using a psychoanalytical approach to bring to light the effects on the individual, we can overcome the negative and catastrophic positions that immobilize human action.

The reading of several contemporary academic references, especially from the last decade, have contributed to thought about climate change, sustainability, and have formed various support networks. These networks offer theoretical and methodological support to different research approaches. At the same time, they try and give support to psychotherapists and other professionals in diverse areas trying to help individuals and communities face the fear and denial of extreme climate events.

Many of the authors analysed, presented significant and challenging approaches. They also offer creativity and innovation, and place individuals and society face-to-face with the true significance of the challenges facing the planet. They affirm that humanity has responsibility in creating climate change, and therefore should be held accountable for creating and implementing sustainable action.

Human responsibility in creating unsustainable projects, combined with the vulnerability of countless regions, affirms that quality of life requires sustainable development and respect for human rights. It also shows the role strategies that have been put in place, have in mitigating and in creating tools for adapting to climate change.

Once again, it is important to emphasize that these approaches and perspectives involve both objective and subjective ways to deal with today's challenges and environmental dilemmas in a very concrete manner. They also offer proposals and mechanisms for finding solutions to environmental problems when sustainability is the goal.

We hope that our work has contributed to broader reflection and that the questions analysed support Quality of Life, recognize the diversity of the planet, show a commitment to present and future generations, and help overcome what appears to be a dystopian reality. We also hope that it allows us to seek the great utopian ideas that nourish us and create hope for better times.

Reference

Mbembe, A. (2020). *O direito universal à respiração*. Available: http://www.ihu.unisinos.br/78-news/598111-the-universal-right-to-breath-article-de-achille-mbembe

Index[1]

A
Access to basic goods, 76
A Common Future, 74
Actors, 73
Africa, 43
Agendas for society and its civilizational process, 3
Agenda 2030, 3, 26, 73, 79, 84, 86
Alleviate poverty, 11
Alternative to development project, 72
Anger, 61
Annual Conference of the Psychiatric Union in London, UK, 42
Anthropocene, 40
Anthropogenic (human caused), 2
Anthropogenic climate change, 53
Anthropogenic global warming (AGW), 21
Anthropogenic warming, 7
Antidepressants and anti-anxiety medications, 22
Anxiety, 52, 58, 62
Apathy, 61, 62
Apocalyptic, 26
Asia, 43
Australia, 43

B
Behaviour towards lower energy demand, 11
Better times, 88
Biodiversity, 10
Bird flu, 29
Black community, 3
Brazil, 43
Brazil, post 2018, 84
Brazilian scientists, 2
Buildings, 11

C
Cambridgecarbonfootprint.org (CCF), 60–62

[1] Note: Page numbers followed by 'n' refer to notes.

Cambridge Sustainable Food, 60
Capacity for disaster preparedness and response, 11
Capitalism and globalization, 40
Capitalist Society, 70
Catastrophe of the future of humanity, 55
Catastrophic climate change, 36
Catastrophism, 35
Challenge, 62
Challenge cultural norms, 58
The challenges of globalization and associated risks, 44
Changes in lifestyle, 11
China, 29
Civil society, 2
Clean water and sanitation, 11
Climate change, 2, 6, 7, 20, 24, 29, 37–39, 52, 55, 56, 61, 62, 76, 85, 87
Climate effects, 8
Climate emergency, 73, 84, 86
Climate Psychology, 56
Climate Psychology Alliance (CPA), 56–59
Coastal ecosystems, 10
Coastal livelihoods, 10
Collective and individual responses, 18
Collective social problems, 23
Colonialism, 72
Common future, 73, 80
Common Future: Inter-social environmental relationships and their challenges, 70
Complex individuals, 61
Complexity of our responses, 61
Concept of development, 73
The concept of subjectivity, 18
Concern, 52
Confusion, 61
Copenhagen (2009), 31
COP 26, 29n3
COP21, 28n2
Corona virus, 29
Covid-19, 2, 73, 76, 83
Covid-19 pandemic, 3, 29n3, 70, 84, 86
Creating climate change, 85
Crisis of biodiversity, 73
Cruellest example, 84
Cultural, 2, 62
Culture, 11

D
Dangerous climate change, 9
Decarbonization, 26
December 2019, 29
Deep Adaptation, 57
Deeper understanding of climate change, 54
Defence mechanisms, 47
Degradation of natural resources, 84
Degradation of socioecological landscapes, 46
Democrat Biden, Joe, 31
Democratisation, 2
Denial, 61, 62
Depoliticalization, 38
Despair, 62
Difficult emotional experiences, 62
The dignity of life, 3, 27
The dilemma between climate and humanity, 40
Disbelief, 62
Disrupted livelihoods, 10
Diverse ecosystems, 62
Diverse impacts on the human psyche, 62
Domestic work, 75
Dominant species, 40
Doubt, 61

E
Ebola, 29
Ecological crises, 9
Ecological damage, 50
Ecological debt, 62
Ecological risks, 37, 52
Ecology, 73
Effective environmental policy, 11
Endless growth, 59
Energy efficiency, 11
Environmental change, 2, 62, 73
Environmental communication, 28
Environmental crisis, 52
Environmental degradation, 85
Environmental development, 2
Environmental dilemmas, 50
Environmental discourses, 50
Environmental melancholy, 50
Environmental politics, 37
Environmental problems, 28
Environmental questions, 62
The European Union, 43
Existential shame, 58
Exploits resources, 26
Extinction Rebellion, 57
Extreme heat, 10
Extreme weather events, 10

F
Face masks, 3
False comfort, 26
Fear and denial of extreme climate events, 87
Fear of climate change, 55
Feelings of insignificance, 62
Feminism, 73
First nations, 43, 73
First Nations people, 3
Fishing communities, 10
Food insecurity, 10
The future, 25
Future generations, 25

G
Gaia, 34
Gasoline, 22
Gender inequality, 84
Glasgow, UK, 29n3
Global Agendas, 73, 80, 84
Global Citizens, 44
Global Civilizational Agendas, 74
Global Climate Change, 7, 8, 34
Global collapse, 80
Global inequality, 75
Global North, 72
Global phenomenon, 39
Global risks, 37
Global society, 24
Global warming, 32
Good Life, 73–75
Good living conditions, 71
Gradualism, 35
Great Britain, 43
Great catastrophes, 29
Greenhouse gases, 34
Greening of modernity, 38
Grief, 61, 62
Guilt, 58, 62

H
Habits of consumption, 26
Haifa Bay, 51
The Happiness Tale, 26
Health crisis, 73, 80
To help individuals and communities, 87
H5N1 influenza, 29
Hope and despair, 59
Human actions, 85
Human activities, 9, 21
Human behaviour, 20
Human condition, 6
Humanity, 9
Humanity's critical responsibility, 43
Humanity's linear vision, 41

Human life, 9
Human participation, 58
Human responsibility, 87
Human rights, 2
Human Rights and Rights of Nature, 74
Human sciences, 33
Human subjectivity, 20
Human sustainability, 41
Hunger, 76

LGBTQIA+ people, 3
Life-changing, 61
Life-degrading, 49
Loss, 62
Loss of marine, 10
Loss of terrestrial and inland water ecosystems, 10
Loss to rural livelihoods, 10
Low-carbon alternative energy sources, 47

I
Impacts of climate change on nature, the economy, and human subjectivity, 55
Increase in temperature, 28–29
The industrialized Global North, 72
Industry, 11, 32
Inequality, 39
Infantilization, 42
Inland flooding, 10
Interdisciplinary approach, 2
Intergovernmental Panel on Climate Change (IPCC), 31, 57
IPCC, 10
IPCC reports, 34

M
Managing the planet, 43
Mass media, 37
Materiality, 44
Mental health, 9
To mitigate the conditions fossils fuels impose, 22
Multiplier effect, 9

N
The narratives we live, 48
New technologies, 11, 26
Nightmare, 26

O
Oppression against women, 75

K
Keeping warming below 1.5 °C, 11
Kyoto (1998), 31, 33

P
Pandemic, 2
Paradox, 71
Paragraph 1 of Article XXV, 79
Paralyzed, 49
Paris Accord, 31
Paris Conference (COP 21), 31, 33
Paris 2015, 28n2, 33
Partnership, 78
Patriarchy, 72

L
Land occupation and use, 34
Land use, 11
Large gaps, 16
Large-scale changes to energy systems, 11
LGBTQI+ and others, 73
LGBTQI+ individuals, 75

Peace, 78
People, 77
Personal change, 62
Physical distancing, 3
Planet, 77
The planetary loss of biodiversity, 85
Pluriversal subject, 45
Policy decisions, 62
Political, 38, 62
Political organizations, 32
Political reality, 62
Politics, 39
Politics of extractivism, 71
Pollution, 6
Poor countries, 71
Poorer populations, 10
Power of science, 35
Predatory relationship, 72
Prejudice, 84
Processing feelings, 62
Pro-environmental behaviour, 48
Prosperity, 41, 78
Psychoanalysis, 62, 73
Psychoanalysis of human subjectivity, 2
Psychoanalytic approach, 24
The psychoanalytic construction of subjectivity, 25
Psychoanalytic territory, 20
Psychological aspects of climate change, 38
Psychological defence, 26
Psychology of sustainable consumption, 48
Psycho-social and ecological issues, 45
Psychotherapists and other professionals in diverse areas, 87
Public pressure on governments, 9
Public response, 9

Q
Quality of life, 11, 25, 62, 71, 73

Quality of Life Concept (QoL), 4–6
Quality of life in a climate change context, 6–11
Quilombos, 3

R
Racism, 84
Recycling, 11
Reduce climate disturbances, 44
Reduced agricultural productivity, 10
Reductions in CO_2 emissions, 11
Reflexive modernization, 36
Regions vulnerable to environmental threats, 11
Regressive moments, 45
Re-use, 11
Rich counties, 71
Rio (1992), 31
Risk of mortality and morbidity, 10
Risks and impacts, 23
Risk society, 37
Risks of severe ill-health, 10

S
Sadness, 52
Safe environment for everyone, 3
São Paulo, Brazil, 45
Sars-COV-2, 29
Scepticism, 34
Scientific discoveries, 85
Secure essential health care services, 11
Semi-arid regions, 10
Shame, 58, 62
Shock, 62
Small steps, 26
Social, 2
Social inequality, 26, 84
Social isolation, 2
Social movement activity, 9
Social natural catastrophe, 71

Social practices, 30
Social processes, 62
Social scientists, 3
Social systems, 62
Sociological, 38
Solidarity, 4
South, 72
Subjective dimension, 27
Subjective experience, 61
Subjectivity, 2, 4, 17, 19, 44, 73, 84
Subjectivity and climate change, 70
Subjectivity and in interpersonal relationships, 72
Sustainability, 2, 45
Sustainability work, 46
Systemic ruptures, 9

T
Traditional populations, 3
Transport mode, 11
Treatment technologies, 11
21st century agenda, 34
Twenty-first century, 85
2021, 31

U
Unconscious masochism, 54
Understanding and support, 59

United Nations (UN), 31, 33
The United States, 43
Universal Declaration of Human Rights (UDHR), 3, 73, 79, 84, 86
The Universal Declaration of Human Rights and Agenda 2030, 74
Unpaid jobs, 75
Unstable experimentation, 44
Urban and rural, 10
Urban populations, 10

V
Vaccination of their populations, 3
Vaccines, 3
Vulnerable, 76

W
Waste, 11
Whether social or individual, 27
Whole self, 61
Women, 2, 3, 73
Women and girls, 75
Wuhan, 29

X
Xenophobia, 84

The manufacturer's authorised representative in the EU is Springer
Nature Customer Service Centre GmbH, Europaplatz 3, 69115 Heidelberg,
Germany. If you have any concerns regarding our products, please
contact ProductSafety@springernature.com

Printed and bound by CPI Group (UK) Ltd, Croydon, CR0 4YY

25/03/2026

02078175-0011